Beautiful China 美丽中国 3

# Public and Commercial Landscape

# 公共商业景观

佳图文化 编

华南理工大学出版社
·广州·

图书在版编目（CIP）数据

美丽中国 3：公共商业景观 / 佳图文化编. —广州：华南理工大学出版社，2013.10
ISBN 978-7-5623-3985-4

Ⅰ．①美⋯　Ⅱ．①佳⋯　Ⅲ．①商业区－景观设计－中国－现代　Ⅳ．① TU984.13

中国版本图书馆 CIP 数据核字（2013）第 163386 号

美丽中国 3：公共商业景观
佳图文化 编

| 出 版 人： | 韩中伟 |
|---|---|
| 出版发行： | 华南理工大学出版社 |
| | （广州五山华南理工大学 17 号楼，邮编 510640） |
| | http://www.scutpress.com.cn　E-mail: scutc13@scut.edu.cn |
| | 营销部电话：020-87113487　87111048（传真） |
| 策划编辑： | 赖淑华 |
| 责任编辑： | 杨爱民　庄　彦　赖淑华 |
| 印 刷 者： | 广州市中天彩色印刷有限公司 |
| 开　　本： | 1016mm×1370mm　1/16　印张：16 |
| 成品尺寸： | 245mm×325mm |
| 版　　次： | 2013 年 10 月第 1 版　2013 年 10 月第 1 次印刷 |
| 定　　价： | 296.00 元 |

版权所有　盗版必究　　印装差错　负责调换

# Preface | 前言

Landscape is traces of nature and human activity on the earth and natural landscape is a miracle created by nature while humane landscape is drawn down by designer souls. Excellent designers will not debase the nature wonder in their pursuit of soul in pen; instead they will utilize the wonder to create various styles of humane landscape.

The shared goal of landscape in American style, British style or Chinese style is to create a graceful dwelling environment. It refers to aesthetics in visual enjoyment and the harmony between human and human, human and environment. For this reason, it requires more than working on external landscaping; designers shall extend their study further to the inner character and definition of landscape, i.e. presenting landscape ecological character, environmental friendly concept etc. A landscape external decides how many people are attracted to come while its inner character decides how many people will really choose here.

A new concept Beautiful China is firstly presented in 18th CPC National Congress. It covers two connotations: the beauty of national nature environment and the beauty of national spirit and social development. Landscape planning and design of the time has departed from the single pursuit of landscape external, focuses more one the inner character of ecology, environment-friendly etc. On the basis of ensuring the landscape external appearance, designers pay attention to the selecting materials of low energy consumption, high efficiency and ultra environment-friendly, fully following the concept of "construction of ecological civilization in a prominent position" proposed in 18th CPC National Congress. The requirements of landscape planning and design of the time tallies right with the spiritual connotation of Beautiful China. The book has selected projects correspond to the double standards of Beautiful China, offering a feast of visual beauty as well as an ecological feast of spirit to single person, society and nation.

The book has professionally analyzed the selected project in multiple perspectives. In the content scheduling, the key points, highlights, design concepts etc are analyzed with a great quantity of professional technical drawings, detailed and informative. We expect to help designers and landscape practitioners create more beautiful landscapes in China, achieve Chinese dream of jointly constructing beautiful China and enduring development.

景观,是大自然和人类活动在大地上的烙印。自然景观是大自然创造出来的神奇,人文景观是设计师笔下的灵魂。好的设计师在追求笔下灵魂的时候不仅不会用手中的笔去掉大自然的神奇,而且还会充分利用大自然的神奇创造出各种风格的人文景观。

但不管是美式风格、英伦风格还是中式风格的景观,共同的目标都是创造一个美丽的栖居环境。这里所谓的"美丽",一是指人视觉上美的享受;二是指人精神上、人与人之间、人与环境之间的和谐之美。为了实现这个目标,纯粹从景观外在美化上下功夫是远远不够的,设计师的笔触应该更深入地伸向景观内在属性的定义上,比如赋予景观生态特质、环保理念等。景观的外在决定了能吸引到多少人,而景观的内在属性却决定了真正有多少人选择这里。

中共十八大首次提出"美丽中国"的新概念。"美丽中国"意涵着两个价值维度:一是国家自然环境之美;二是国家精神及其社会发展之美。当代景观规划的设计已经脱离了纯粹的景观外在化追求,而更加注重生态、环保等景观的内在属性。设计师在保证景观外在美观的基础上,注重选用低能耗、高效能、超环保的材料,全面贯彻中共十八大"把生态文明建设放在突出的位置"的精神理念。当代景观规划设计的要求正契合了"美丽中国"的精神内涵。本书在案例选择上执行的就是"美丽中国"的双重标准,既给人带来一次美的视觉盛宴,又让个人、社会和国家在精神上享受一场生态盛飨。

本书从多角度详细且专业地分析了"美丽"的景观案例。内容编排上,分别从景观案例的关键点、亮点、设计思想等方面入手,配合大量的专业技术图纸,资料丰富而详实。我们的努力是为了让设计师及景观从业者在中国的蓝图上创作出更多更美的景观,在景观设计上共同实现建设美丽中国、实现中华民族永续发展的中国梦。

# CONTENTS 目录

## Sci-Tech Park Landscape 科技园区景观

Factory Complex of Nanjing Estun Automation Technology Co., Ltd. ——— 002
南京埃斯顿自动控制技术有限公司厂区

Ningbo Chunxiao Pioneering Service Center ——— 008
宁波春晓创业服务中心

Wuhan Future City ——— 014
武汉未来科技城

Guangxi Egret Park ——— 022
广西白鹭公园

Xuzhuang Software Park ——— 034
徐庄软件园

## Commercial Landscape 商业景观

Landscape Design for Air City, Beijing ——— 044
北京顺义中建翼之城商业景观设计

Changsha Taskin City Landscape Design ——— 056
长沙德思勤城市广场景观设计

Suzhou Harmony Times Square ——— 064
苏州工业园圆融时代广场

Chongming Yulin Business square ——— 072
崇明育麟商务广场

## Tourism Landscape 旅游度假区景观

The Overall Planning of Taizhou Xianju Tea Valley Resort ——— 082
台州仙居茶溪谷度假区总体规划

Shanxi Zhangbi Castle SPA Resort ——— 086
山西张壁古堡温泉养生度假区

Kashi North Lake Ecological Park Landscape Planning and Design ——— 092
喀什北湖生态公园景观规划概念设计

Dalian Software Park Donggou Mountain Park ——— 110
大连软件园东沟山体公园

Santorini Hotel Landscape Conceptual Design ——— 118
圣托里尼酒店景观概念设计

Planning of Nanshufeng King of Medicine Health Preservation Valley ——— 124
南树峰药王养生谷

| | |
|---|---|
| **The Concept Planning of Yangjiang Spring Resort** | 128 |
| 阳江温泉度假村 | |
| **Shenzhen Aegea · Holiday Mansion** | 134 |
| 深圳爱琴海·假日公馆 | |
| **Hefei Lakefront Forest Park** | 142 |
| 合肥滨湖森林公园 | |
| **Taizhou OCT Hot Spring SPA Landscape** | 150 |
| 泰州华侨城温泉 SPA 景观 | |

## Urban Planning 城市规划

| | |
|---|---|
| **Kunlun Rainbow Landscape Design in Hecheng Town, Jiangmen City** | 164 |
| 江门市鹤城镇昆仑彩虹景观设计 | |
| **Luzhou City Xueshishan Park** | 174 |
| 泸州市学士山公园 | |
| **Jiuhuashan Animation Theme Park** | 182 |
| 九华山动漫主题乐园 | |
| **China (Chongqing) International Garden Expo** | 188 |
| 中国（重庆）国际园林博览园 | |

## Public Landscape 公共景观

| | |
|---|---|
| **South Bank Riverside Scenic Zone of Zhengshui River, Hengyang** | 198 |
| 衡阳蒸水河南岸滨江风光带 | |
| **Hebei Qian'an City Huangtaishan Park** | 202 |
| 河北省迁安市黄台山公园 | |
| **Linyi International Sculpture Park** | 208 |
| 临沂国际雕塑公园 | |
| **Landscape Project of Huangshi Magnetic Lake** | 218 |
| 黄石磁湖湖景工程 | |
| **Shenzhen OCT East** | 230 |
| 深圳东部华侨城 | |
| **Shenzhen Shekou Taizi Road** | 238 |
| 深圳蛇口太子路 | |
| **Tianjin Cultural Park** | 244 |
| 天津文化中心 | |

# Sci-Tech Park Landscape
# 科技园区景观

Eco Measure
生态措施

Simple Appearance
简约形态

Space Sequence
空间序列

# SCI-TECH PARK LANDSCAPE
# 科技园区景观

## MODERN SIMPLE STYLE
## 现代简约风格

**KEY WORDS** 关键词

### THEME FEATURES
### 主题特色

### HYDROPHILIC CHARACTER
### 亲水特性

### LANDSCAPE NODE
### 景观节点

Location: Nanjing, Jiangsu
Developer: Nanjing Estun Automation Technology Co., Ltd.
Landscape Design: A&I International
Chief Designer: Cao Yuying
Design Team: Lian Jiajia, Sun Jie
Land Area: 43,324 m²
Total Floor Area: 90,000 m²

项目地点：江苏省南京市
开 发 商：南京埃斯顿自动控制技术有限公司
景观设计：安道国际
主创设计师：曹宇英
设计团队：连佳佳　孙杰
建筑用地：43 324 m²
总建筑面积：90 000 m²

# Factory Complex of Nanjing Estun Automation Technology Co., Ltd.

# 南京埃斯顿自动控制技术有限公司厂区

## FEATURES 项目亮点

The design in modern, simple and novel language builds an architectural image with strong style characteristic and appeal, and a graceful garden environment.

通过现代、简洁且新颖的设计语言，营造一个风格特征强烈、具有感染力的建筑形象和优美的园区环境。

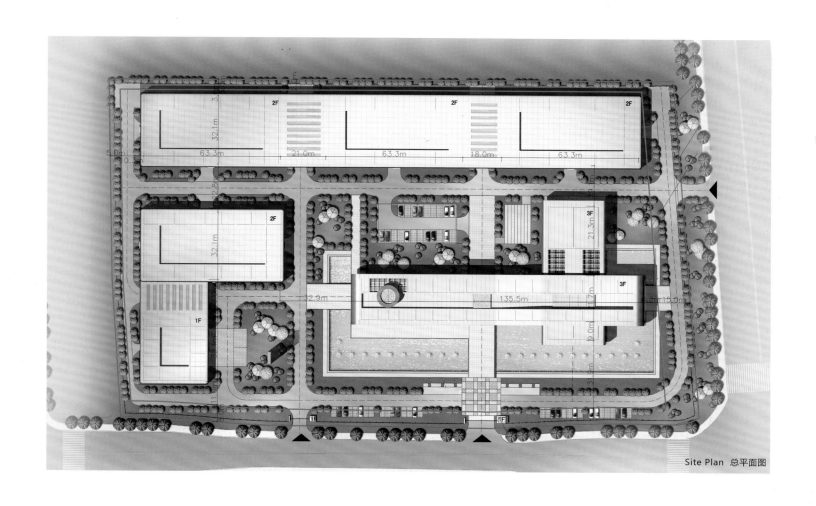

Site Plan 总平面图

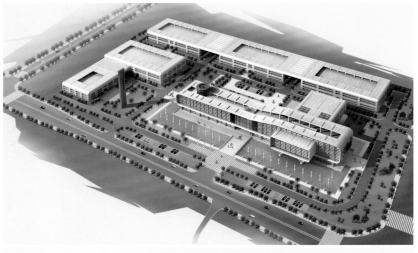

## Overview 项目概况

The project land is a plot of industrial land located in Jiangning District, Nanjing, and east to Shuige Road, south to Huayuan Street. It is 260 m long in west and east direction, 160 m in north and south direction, shaping in a rectangle with flat terrain. It connects to urban roads in east and south sides, adjacent to a road across a river, convenient for traffic.

本地块位于南京市江宁区，东至水阁路、南至花园街，用地性质为工业用地。地块东西长约260 m，南北长约160 m，近长方形用地，现状地形平坦。地块东、南边与城市道路相接，西侧隔小河道与道路相邻，交通便利。

## Design Goal 设计目标

The design adopts natural way to arrange landscapes. Based on satisfying the production and office demands of modern enterprises, targeted at building a comfortable and graceful modern industrial park, the design in modern, simple and novel language builds an architectural image with strong style characteristic and appeal, and a graceful garden environment.

设计以自然的手法来布置景观。以满足现代工业企业生产办公需求为基础，以创造舒适优美的现代工业园区为目标，通过现代、简洁且新颖的设计语言，营造一个风格特征强烈、具有感染力的建筑形象和优美的园区环境。

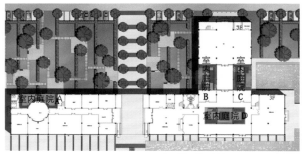

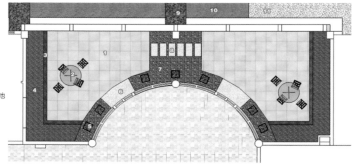

室内庭院 A 平面图
1、休憩平台
2、现代休憩坐椅
3、散置黑色卵石收边
4、整形灌木
5、可移动式现代不锈钢树池
6、整石平桥
7、散置黑色卵石铺地
8、整石汀步
9、室外整形灌木
10、室外大草坪
11、室外白色小碎石

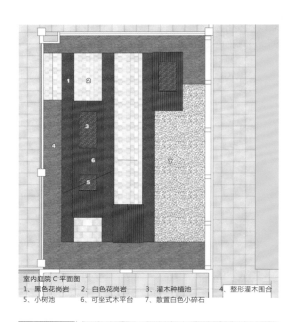

室内庭院 C 平面图
1、黑色花岗岩　2、白色花岗岩　3、灌木种植池　4、整形灌木围台
5、小树池　6、可坐式木平台　7、散置白色小碎石

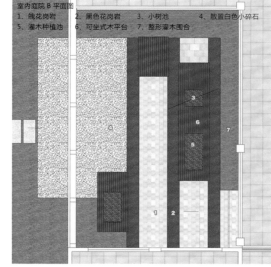

室内庭院 B 平面图
1、魄花岗岩　2、黑色花岗岩　3、小树池　4、散置白色小碎石
5、灌木种植池　6、可坐式木平台　7、整形灌木围台

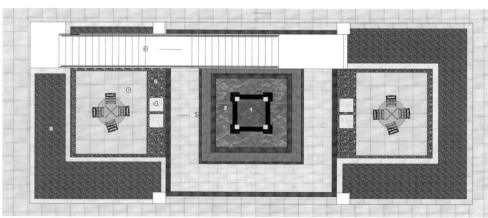

室内庭院 D 平面图
1、特色跌水　2、蓄水池　3、楼梯（原有）　4、整石汀步
5、散置黑色卵石　6、休憩平台　7、下沉式特色铺装　8、花坛

### Landscape Node 景观节点

The project planning has presented a modern factory area with theme and characteristic, through the comprehensive process of architectures, greening, roads, landmarks, and iconic articles. The large-scale waterscape in the south side effectively keeps away the interference of road noise, enhances the internal environment quality and forms a large-scale central green land between office complex building and workshops. A small green land is shaped in west side additionally, and the urban greening belt in east side has been taken into the block; greening in various zones has strengthened the landscape structure of the entire area, graceful and resources balancing.

The major landscape axis is in west and east direction is formed with the waterscape in south side of the project area. With the green land between office building and workshops as the sub-axis, it connects each plot of green land, building an intertwined greening system with landscapes and roads.

Articles with cultural characters are set in proper positions to form humane landscape, improve enterprise cultural quality.

Landscape water system is processed with economical, ecological and environment-friendly technology, with great realistic possibility. The design highlights landscape hydrophilicity and accessibility under the requirements of being clean and beautiful.

项目规划通过建筑、绿化、道路、地形以及标识性小品的综合处理，构筑一个具有主题和特色的现代厂区。地块南侧的大面积水景观有效隔离道路噪音的干扰，并对厂区的环境品质起到巨大的提升作用；在办公综合楼和车间中心形成较大规模的中心绿地，另在地块西侧形成一个小型绿地，地块东侧的城市绿化带也为区块所使用；各个区域的绿地强化了整个厂区优美均好的景观结构。

该项目的景观轴线是以厂区南侧水景形成东西景观主轴，并以通过办公综合楼和车间之间的绿地为次轴串联各个小型绿地，以景成点、以路成线，形成交织的网络状绿化体系。

项目在适当部位设置具有文化特征的小品，形成具有陶冶情操、修身养性功能的人文景观，有力提升企业的文化品质与底蕴。

景观水系采用经济、生态、环保的水处理技术，具有很强的现实可能性。同时，设计强调在满足洁净和美观的要求下，通过一定的设计手段实现景观的亲水性和可及性。

# SCI-TECH PARK LANDSCAPE
科技园区景观

## MODERN SIMPLE STYLE
现代简约风格

### KEY WORDS 关键词

**IMAGE SPACE**
意象空间

**ARTISTIC TEXTURE**
艺术肌理

**LANDSCAPE NODE**
景观节点

Location: Ningbo, Zhejiang
Developer: Ningbo Chunxiao Development & Construction Co., Ltd.
Landscape Design: A&I International
Chief Designer: Zhao Difeng
Design Team: Yu Youxian, Jiao Qinghe
Total Landscape Area: 9,362 m²

项目地点：浙江省宁波市
开 发 商：宁波春晓开发建设有限公司
景观设计：安道国际
首席设计师：赵涤烽
设计团队：余友贤　焦清合
总景观面积：9 362 m²

# Ningbo Chunxiao Pioneering Service Center
宁波春晓创业服务中心

## FEATURES 项目亮点

The designers' interpretation of ideal life environment is conveyed by an interesting image space, an enduring form constructed through language of lines, texture and colors.

通过线条、肌理和色彩的语言构筑一个持久的形式，创造充满趣味性的意象空间，传达设计师对于理想生活环境的理解。

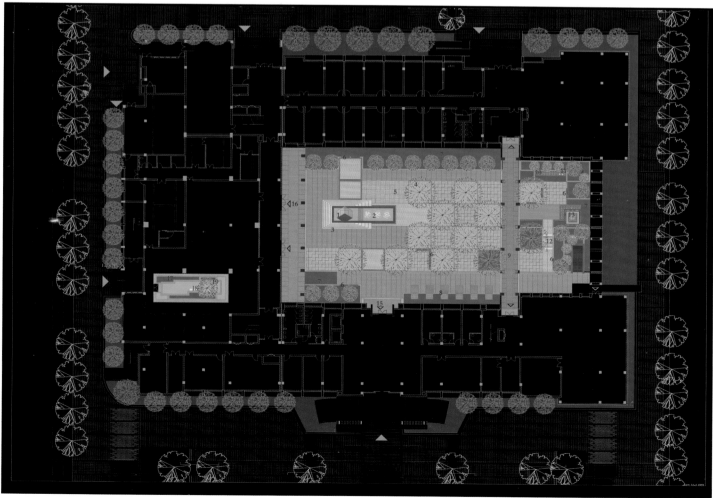

1. 特色正文体雕塑
2. 涌泉
3. 景观台阶
4. 银杏种植
5. 特色拉丝面格子铺
6. 休憩座椅
7. 灌木
8. 特色数字景观雕塑
9. 建筑内廊
10. 散铺鹅卵石
11. 特色灯饰
12. 台地
13. 景观雕塑小水景
14. 地下室入口
15. 创业服务中心景观主入口
16. 餐厅前廊
17. 台地式内庭
18. 景观小雕塑
19. 乔木种植

Site Plan 总平面图

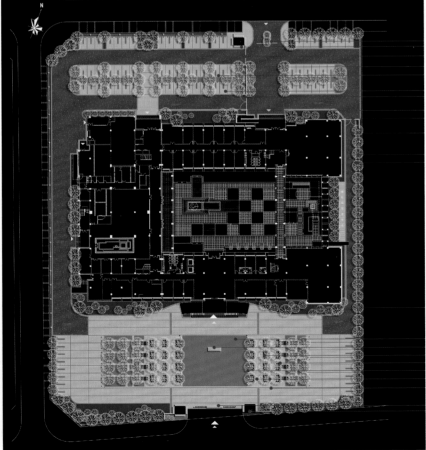

机动车道闸
门卫亭
入口大门
围墙
车挡

停车位
临时消防通道大门
灌木
临时消防通道大门

沥青道路

围墙
灌木

灌木

耐践踏草皮
升旗台
树坑
树阵
草坪灯（带春晓Logo）
人行通道
机动车道

主入口大门
门卫亭

Asphalt Road Square Scheme
沥青道路广场方案

### Overview 项目概况

The project covers the landscape design of plaza and building atrium.

项目景观设计包括广场景观设计和建筑中庭景观设计两个部分。

### Design Thread 设计思路

The designers adopt language of lines, texture and colors to construct an enduring form, to deliver the realization of nature. The visual experience of spatial distance, closure, expansion or shrink is united with dynamic experience which refers to body movements in spaces. Elements are arranged in order to build an interesting image space, delivering designers' interpretation of ideal life environment.

设计师用线条、肌理和色彩的语言来构筑一个持久的形式，传承人们对自然的体悟。人们对空间的距离、封闭、扩展或收缩的视觉体验是与动感体验，即与人们在空间中的移动或穿过空间时的身体动作结合在一起的。在该项目中通过对元素的有序排列，创造充满趣味性的意象空间，传达设计师对于理想生活环境的理解。

## ▶ Landscape Node  景观节点

Architectural elements are utilized in the landscape design of plaza, in pursuit of organic order and balance. The architecture style is extended to the landscape, as a way of circle, bringing sense of movement in the internal site. A great deal of greening design is a way of symbiosis in harmony, with definite sense of power to destruct before construct. It is like a way of penetration in every detail without any erosion; it is an optimization ion level and colors.

In the design of atrium, designers expect to build a space for recreation as well as a place close to nature after work. The interlacement of straight lines and the greening design of height difference produce various space with different scales in the shape of chara for "回".

The design of waterscape and sculpture in atrium integrates with the architecture style, simple and modern. The small drop water landscapes have become the focus in the site, enlivening the space. Large trees arranged elaborately bring the shading effect and construct the visual center, delivering nature touch to people who even work indoor.

广场景观设计以建筑的元素来寻求有机的次序和均匀，将建筑风格延伸至景观之中进行循环，在场地内部的行进中赋予运动感和场所的精神感。场地内大量的绿化设计也是一种协调的共生方式，极具力量感，先打破而后生成。设计师对于绿化的处理手法类似一种对场地的渗透，精细至每个毛孔，同时并没有对场地造成侵蚀，而是一种层次和色彩上的优化。

在中庭的设计中，设计师希望构筑一个在工作之余能够亲近大自然的场所和缓解疲劳的休憩空间。通过直线线条的交错以及绿化的高差设计，产生了多个不同尺度的"回"字形空间，高高低低的错落之中非常具有趣味性。

中庭两处水景和雕塑的设计很好的融合了建筑风格，简约而充满现代感。几处小跌水不仅构成了场所的焦点，同时极好地活跃了空间的气氛。中庭各处巧妙设置的大乔木，不仅形成了林荫效果，也构成了场所的视觉中心，能让处于室内工作的人们抬起头就能感受到自然的气息。

# SCI-TECH PARK LANDSCAPE
科技园区景观

## MODERN STYLE
现代风格

### KEY WORDS 关键词

**GREEN SYSTEM**
绿化系统

**LANDSCAPE NODE**
景观节点

**ECOLOGICAL FUNCTION**
生态特质

Location: Wuhan, Hubei
Landscape Design: Shanghai WEME Landscape Engineering Co., Ltd.
Total Planning Area: 66,800,000m²

项目地点：湖北省武汉市
景观设计：上海唯美景观工程设计有限公司
总规划面积：66 800 000 m²

# Wuhan Future City
武汉未来科技城

## FEATURES 项目亮点

The landscape design follows the overall planning to insist on four principles to be ecological, open, advanced and scientific; highlight four ideas of "co-existence, green, open and culture".

景观设计遵循规划总体要求，体现生态性、开放性、先进性、科学性四项原则，突出"共生、绿色、开放、人文"四大理念。

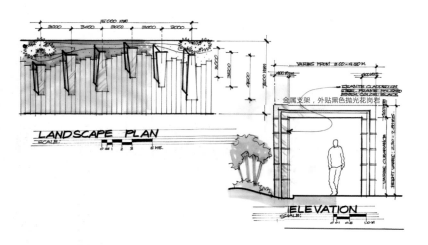

### Overview 项目概况

Future City is located in the National Independent Innovation Demonstration Zone of southeast Wuhan. On the north of Gaoxin Road and on the west of the outer ring road is the 26,000,000 m² starting area where there will be New Energy Research Institute, R&D Center, incubating zone, business area and residential area. As the landscape design for phase I (Zone A) of the starting area this project achieves a plot ratio of 1.13 in this 333,335 m² site. The site is composed of nine plots from A01 to A09, with the surface channel of Longshan Reservoir running through. According to the requirement, the channel area will be transformed to be a wetland park in Longshan Creek area.

武汉未来科技城位于武汉东南部东湖国家自主创新示范区。其中，高新大道以北、外环线以西的 26 000 000 m² 区域为起步区，规划有新能源研究院、研发区、孵化区、商务区及住宅区，其中起步区一期占地约 333 335 m²，在概念设计中称为 A 区。本项目即为 A 区范围内的景观设计，该区容积率为 1.13，园区由 A01～A09 的九个地块项目组成。其中，原龙山水库排水明渠从园区中穿过，要求将排水明渠改造为龙山溪湿地景观公园。

## Design Concept 设计理念

The landscape concept for the sci-tech park: "fast work and slow life for a new future city"; creating a high-tech park in the wetland area with both the international standard and local characteristics. The landscape design follows the overall planning to insist on four principles to be ecological, open, advanced and scientific; highlight four ideas of "co-existence, green, open and culture". From ground to roof, it creates a three-dimensional landscape system for a world-class new city.

科技园区的景观设计理念定为："快工作、慢生活，打造科技未来新概念"；打造具有国际标准的、同时又具有本地特色的湿地景观与高科技园地环境。景观设计遵循规划总体要求：体现生态性、开放性、先进性、科学性四项原则，突出"共生、绿色、开放、人文"四大理念，从地面景观到屋顶绿化，打造立体的城市绿化系统，建成国内外一流的园林城市。

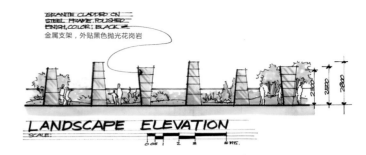

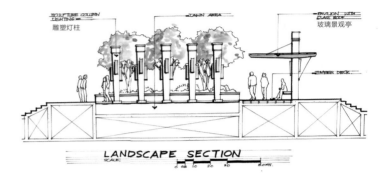

## Landscape Node 景观节点

The sci-tech park is composed of "one belt and one park, three longitudinal axes and four horizontal axes, the fifth facade, and nine zones". Different functions are well arranged to create an ecological and harmonious city; a green and low-carbon city; an open and sharing city; a cultural and innovative city. During the process of design and practice, new technologies, new materials and new techniques are applied to highlight the feature of innovation. In addition, there is also bionic architecture here. Moreover, the fifth facade-roof garden is well designed to be ecological.

In the wetland park, lakes of different water-levels are designed to form beautiful cascades. Together with different plants, it becomes the focus of the whole park.

科技园区由"一带一园、三纵四横、第五立面、九大分区"组成。功能分区旨在打造生态、和谐之城，绿色、低碳之城，开放、共享之城，人文、创新之城。在设计与实施的过程中，加强了高新技术、新材料、新工艺的运用，体现了景观设计新颖性的特点。另外还在园区设计了仿生建筑，并对第五立面——屋顶花园进行了生态设计，体现了景观的生态特质。

项目内的湿地公园由不同标高的水面组成大小不同的湖面，分旱、雨季等不同的季节，形成"泉水叮咚"与"叠水"的效果，并配置相应植物，从而形成园区的亮点。

# SCI-TECH PARK LANDSCAPE
# 科技园区景观

## MODERN STYLE
## 现代风格

### KEY WORDS 关键词

**ECO NATURE**
生态特质

**LANDSCAPE NODE**
景观节点

**CULTURAL ATMOSPHERE**
人文氛围

Location: Fangchenggang, Guangxi
Developer: Guangxi Fangchenggang Housing and Urban-Rural Planning Committee
Landscape Design: Shenzhen L&D Architecture and Landscape Design Co., Ltd.
Land Area: 269,000 m²
Landscape Area: 123,000 m²

项目地点：广西防城港市
开 发 商：广西防城港市住房与城乡规划委员会
景观设计：深圳灵顿建筑景观设计有限公司
用地面积：269 000 m²
景观面积：123 000 m²

# Guangxi Egret Park
# 广西白鹭公园

## FEATURES 项目亮点

Based on the surrounding landscapes, sea and mountains, the designers pursue a simple and unified style in the overall design to set up new image and new landscape for the city.

设计依托当地的周边景观，以海水为源、以山体为依托、以水为景，整体设计追求风格上的简练、统一。

### ▶ Overview 项目概况

The project is located in Fangchenggang City which is at the southwest end of the Chinese Mainland coastline, north on Nanning City, south to the Beibu Gulf, east to QinZhou City and close to Vietnam in the west. It belongs to the Subtropical Monsoon Zone with adequate sunshine and convenient traffic system, not far from the municipal government. The whole project occupies a land area of 269,000 m² and a landscape area of 123,000 m².

项目所在的广西防城港市，地处中国内地海岸线西南端，北连南宁市，南临北部湾，东接钦州市，西邻越南，属于亚热带季风带，阳光充足，周边的交通较为便利，市政府也距离不远，环境良好。整个项目用地面积为26.9万m²，景观设计面积12.3万m²。

Site Plan 总平面图

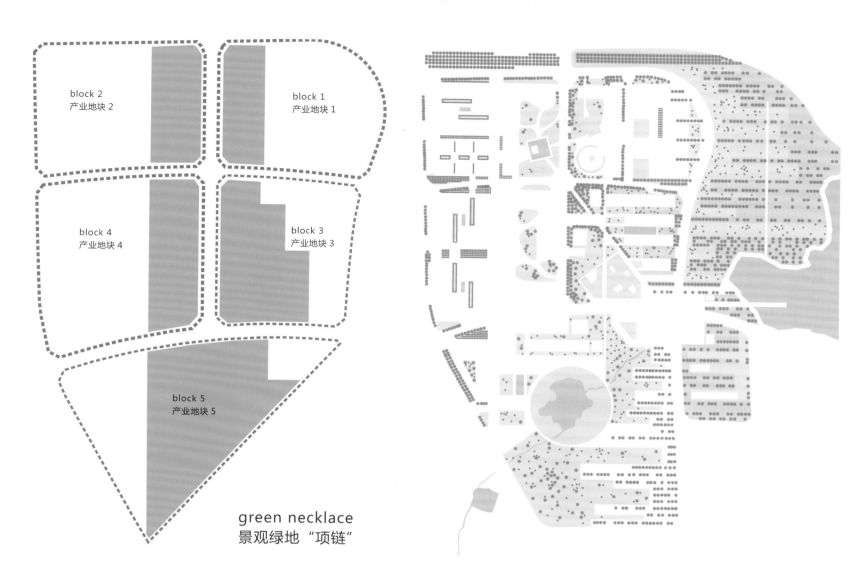

green necklace
景观绿地"项链"

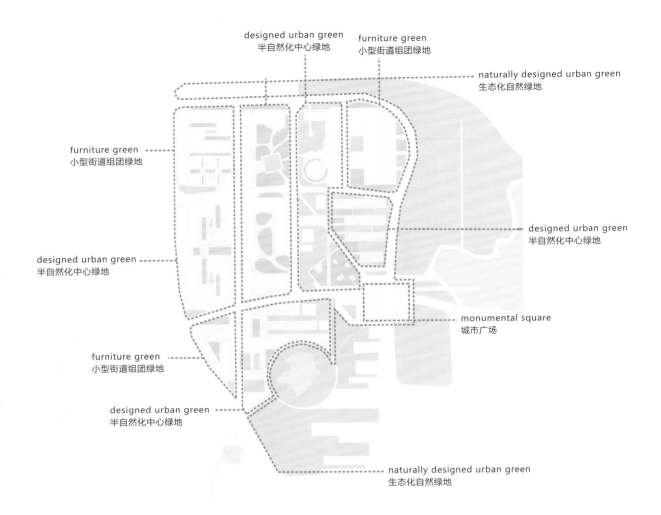

### Design Concept 设计理念

The project bears a complicated topography of mountainous terrain and gets close to the sea; there is also a large area of fresh water between mountains and municipal planning area around the park. There is irregular distribution of waters on the project base and waters in the main entrance plaza build landscape water system. The north of the Xinggang Road near the sea uses the local labor resources to turn itself into a recreational swamp area so as to reduce the environmental damage and get natural & ecological balance.

Based on such a complex mountain terrain, the design tries a simple design method to emphasize the modernity and symbolism, thus reflecting the characteristics of the park. Relying on the local surrounding landscape, sea and mountains, the designers pursue a simple and unified style in the overall design that adopts classified design to make the project a higher quality of the environmental and human landscape, sticking to the people-oriented principle to set up new image and new landscape for the city.

项目的地形是较为复杂的山地地形，项目位置靠近海，山与山之间形成了大面积的淡水区域，公园周边为市政规划区域。基地内有不规则的水域分布，主入口广场的水域使设计师将两个水域相连，营造景观水系。兴港大道北侧靠海，利用当地劳动资源设计成可游玩沼泽区域，最大化地减少破坏，保护自然生态平衡。

面对复杂的山体地形，在设计上尝试用简约的方法来强调时代感和标志性，从而体现这个时代公园应有的特色。依托当地的周边景观，以海水为源，以山体为依托，以水为景，整体设计追求风格上的简练、统一，各个环节分类设计的方法使得项目拥有更高品质的环境景观和人文景观，坚持以人为本，为城市树立新形象、新景观。

# SCI-TECH PARK LANDSCAPE
# 科技园区景观

## MODERN SIMPLE STYLE
## 现代简约风格

### KEY WORDS 关键词

- LANDSCAPE NODE 景观节点
- LANDSCAPE TONE 景观基调
- ECO & ENVIRONMENT FRIENDLY 生态环保

Location: Nanjing, Jiangsu
Client: Management Committee of Nanjing Xuzhuang Software Industrial Base
Landscape Design: Shenzhen DongDa Landscape Design Co., Ltd.
Total Planning Area: 150,000m²

项目地点：江苏省南京市
委托单位：南京徐庄软件产业基地管理委员会
景观设计：深圳市东大景观设计有限公司
总规划面积：150 000 m²

# Xuzhuang Software Park
# 徐庄软件园

## FEATURES 项目亮点

In pursuit of harmony, perfection and the balance between cost saving and architectural effect, designers use concise, pure and natural expression technique, combine topography and function, creating a simple, clean, natural, elegant and generous landscape tone.

设计师采用简约、纯净、自然的表现手法，追求和谐与完美，讲求经济节约与效果的平衡，并结合地形和功能，奠定了简洁明快、自然优雅、大气沉稳的景观基调。

### Overview 项目概况

Located in the center of scenic area of Nanjing Xuzhuang Software Park, this project is designed to serve as a large open public waterfront park which will enhance the economic value of the land and arouse the attention on ecological environment.

基地位于南京徐庄软件园的中心景区,被定位为公共开放空间的大型水岸公园,它将提升地块的经济价值,也将唤起人类对生态环境的重视。

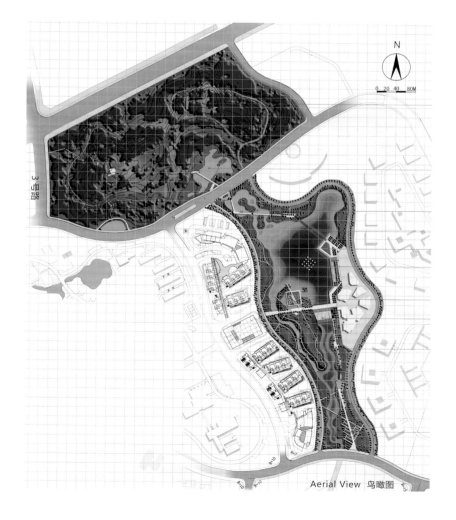

Aerial View 鸟瞰图

### Design Concept 设计理念

In pursuit of harmony, perfection and the balance between cost saving and architectural effect, designers use concise, pure and natural expression technique, combine topography and function, creating a simple, clean, natural, elegant and generous landscape tone. In accordance with various requirements, rational analysis and logical design are used, and clear idea & thought are unfolded in the concise landscape form. Lake, shoreline, terrain, bridge and gallery frame show a very uniform, rational, simple and beautiful shape of this project, in addition, the keyboard-like seats and avenue of stars reveal its unique content and keep it interesting.

设计师采用简约、纯净、自然的表现手法，追求和谐与完美，讲求经济节约与效果的平衡，并结合地形和功能，奠定了简洁明快、自然优雅、大气沉稳的景观基调。设计中按照各种需求，对功能予以理性的分析，用逻辑和秩序进行设计，在简约的景观形式中展现清晰的观念和思考。湖体、岸线、地形、场地、桥梁、廊架等，呈现出极为统一且理性简约的优美形态，而键盘座椅、星光大道等，又赋予其独特的内涵和趣味性。

## Ecological Measures of Landscape 景观生态措施

Protective measures are improved and enhanced for the northernmost Shizi Mountain. Tour route, footpath and staircase are nestled among hills naturally, and the landscape along both sides is optimized. In addition, landscape nodes are distributed at the entrance and the mountain top, modern techniques such as information window and CD-ROM pavement are largely used to adorn the park.

The most prominent is water treatment, including the design of wetland and the planting of aquatic plants which keeps a long-term self-cleaning condition for the water. Moreover, reclaimed wastewater treatment system is also designed to further enhance ecological idea and improve the technical content of this project.

最北边的狮子山是在保护原始生态林区的基础上进行完善和提升，依山就势设置上山游览路线，以自然方式设置步道、阶梯，并对道路两侧作重点的景观优化。同时在入口区域及山顶等处设景观节点，穿插现代手法，点缀场地、设施，以及如信息窗、光碟铺装等具有软件园特色的主题小品。

项目尤为突出的是成功的生态水处理设计及效果，包括设计湿地、栽植多种类水生植物，使水体得到生态保持而长期保持自我清洁的较稳定状态，另外设计了"中水处理系统"进一步提升生态环保理念，也提高了项目的技术含量。

# Commercial Landscape
# 商业景观

**High Fashion**
时尚高端

**Recreational Facilities**
休闲配套

**Landscape Space**
观赏空间

# COMMERCIAL LANDSCAPE
商业景观

## MODERN STYLE
现代风格

### KEY WORDS 关键词

**STREAMLINE DESIGN**
流线设计

**BUSINESS CHARACTERISTIC**
商业特质

**LANDSCAPE NODE**
景观节点

Location: Shunyi District, Beijing
Landscape Design: DDON Associates
Land Area: 100,000 m²

项目地点：北京市顺义区
景观设计：笛东联合（北京）规划设计顾问有限公司
占地面积：100 000 m²

# Landscape Design for Air City, Beijing
北京顺义中建翼之城
商业景观设计

## FEATURES 项目亮点

The design starts from the overall plan and creates an open and dynamic landscape space according to different functions and requirements.

设计从整体园区的景观规划出发，根据商业空间的性质以及需求，打造了一个开放而富有活力的景观空间。

## Overview 项目概况

Located in Beijing Airport Economy Zone, at the southwest of Air CBD, the project is about 3km from T3 terminal and it takes only 45 minutes by car to Chaoyang CBD. Occupying a land area of 100,000m², it becomes an international CBD for air transportation, aviation training, finance and insurance as well as the supporting business services, catering and entertainment.

项目位于北京临空经济开发区，处在国门商务区西南部，距T3航站楼约3 km，距朝阳CBD区域仅45分钟车程。项目占地面积100 000 m²，重点发展航空运输、航空培训、金融保险等生产性服务业，配套发展商务服务、餐饮娱乐等现代服务业，成为现代化国际商务区。

## Design Idea 景观定位

The landscape design conforms the overall planning and matches the architectural style to provide modern landscape spaces. It maximizes the value of the CBD and tries to meet people's requirements for high-quality life.

整体园区景观设计遵循园区整体规划定位，遵循整体建筑的风格，打造现代风格特色的景观空间；最大化地体现园区价值，以人的感知为设计导向，满足生活需求及品质要求。

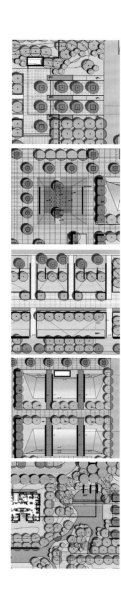
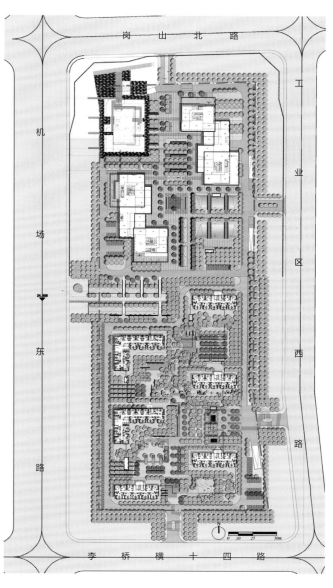

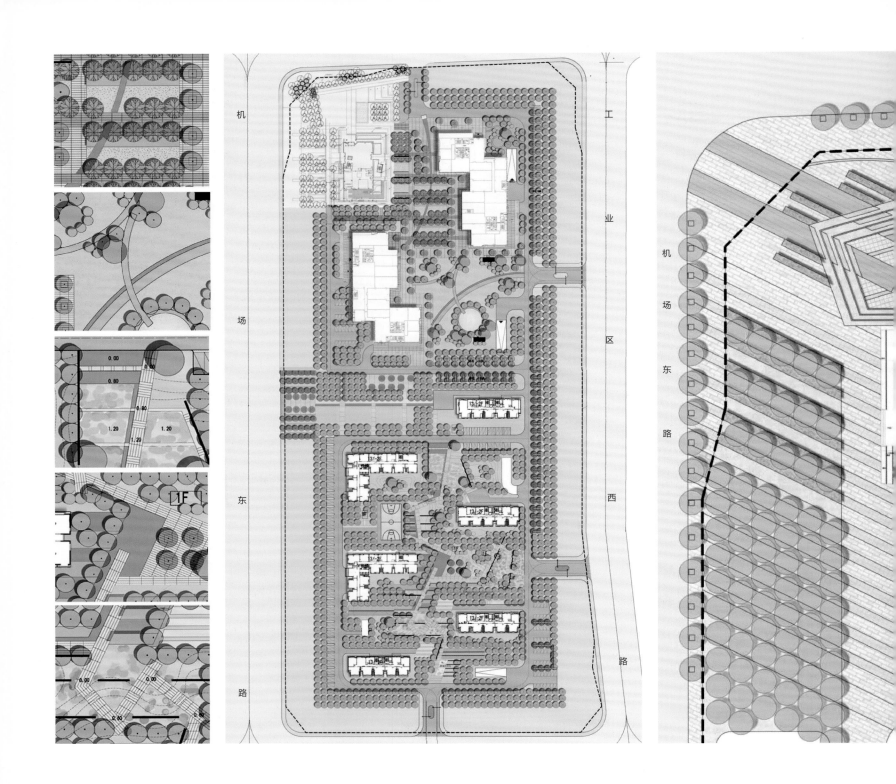

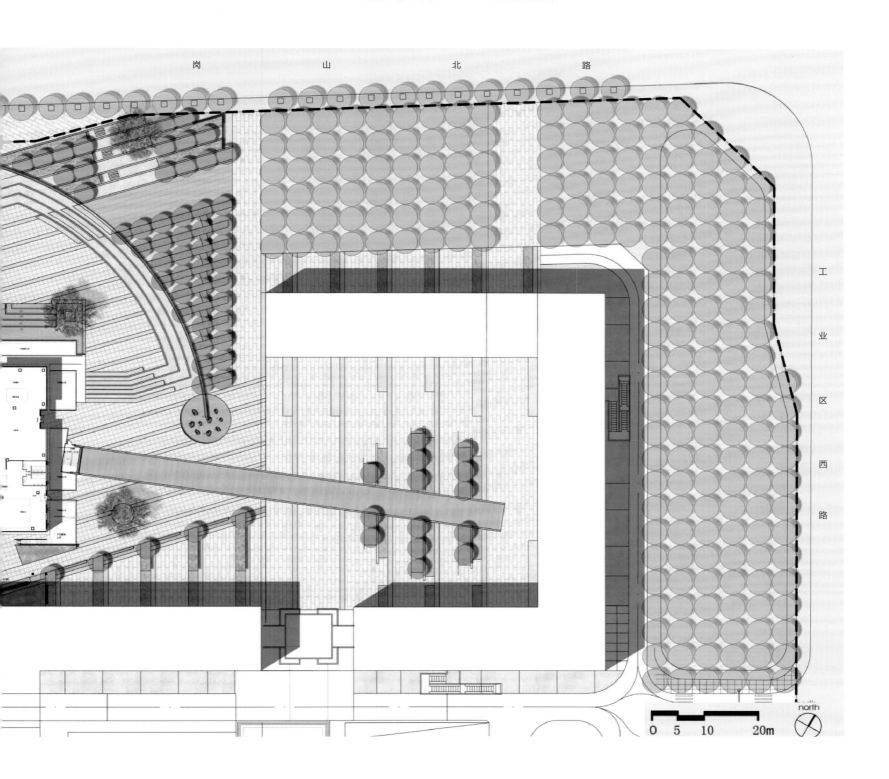

### Landscape Node 景观节点

According to different functions, the business area in the north is divided into three parts, namely, the entrance experience area, central square activity area, and the small stay area. These areas complements each other from scales and functions, creating a harmonious environment. People will be led to the outdoor patio from the entrances on the south and north side. The central patio is sub divided into small spaces, for example, the communication space under trees, the sunken square and the grass slopes. All these spaces will meet people's requirements for visiting, recreation, stay, communication, walk, sports, etc. It will greatly enrich the interface of the whole development and shape a contrast with the exterior space.

The urban square locates between the business area and the residential area as the transition space. The design for this part focus on the view points and leads people to different spaces with the roadside landscapes.

北部商业区根据建筑使用性质及功能需求，共分为三大部分，即：入口体验区、中央广场活动区、小尺度停留区，每个区域在尺度与功能上相互补充，起到良好的协调作用。所有动线通过南北两侧的入口将实线与动线导入室外中庭花园，中庭花园通过空间拆分，形成多种尺度的室外空间：林下停留交往空间，下沉式广场空间，穿行式草坡空间。所有的空间均满足人群对场地的需求，并产生相应的活动行为：观赏、休憩、驻足、交流、散步、运动等。以此带动整体内部界面的丰富性，并使之与外部空间形成反差、对比。

城市广场作为转换空间，介于商业区和住宅区之间。此处的设计初稿就基于这种设计理念，通过路边的景观，将人们带往另一空间。

## COMMERCIAL LANDSCAPE
商业景观

## MODERN STYLE
现代风格

### KEY WORDS 关键词

**THEME CREATION**
主题营造

**AXIS LANDSCAPE**
轴线景观

**LANDSCAPE NODE**
景观节点

Location: Changsha, Hunan
Landscape Design: DDON Associates
Land Area: 374,000 m²

项目地点：湖南省长沙市
景观设计：笛东联合（北京）规划设计顾问有限公司
占地面积：374 000 m²

# Changsha Taskin City Landscape Design

## 长沙德思勤城市广场景观设计

### FEATURES 项目亮点

This case insists in such a design philosophy that given consideration to both natural ecology and entertainment shopping, whose specialty is based on stage and commercial street as two major landscape axis.

项目坚持自然生态与休闲趣味购物兼顾的设计理念，以舞台和商业景观街为主的两大景观轴线是项目的特色。

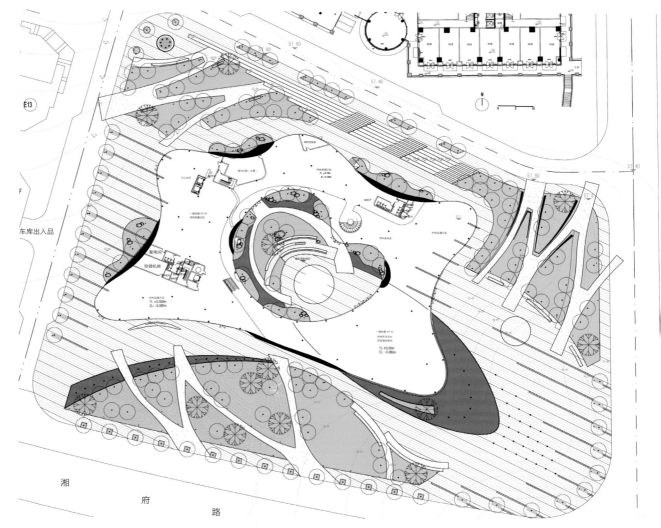

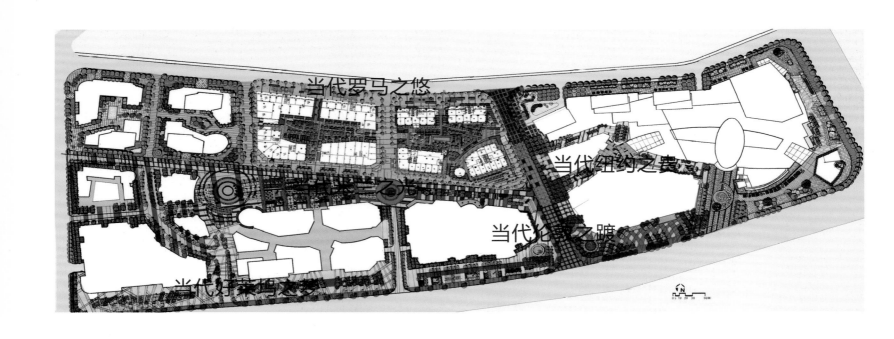

当代罗马之悠

当代纽约之衰

当代比萨之跛

当代好莱坞之衰

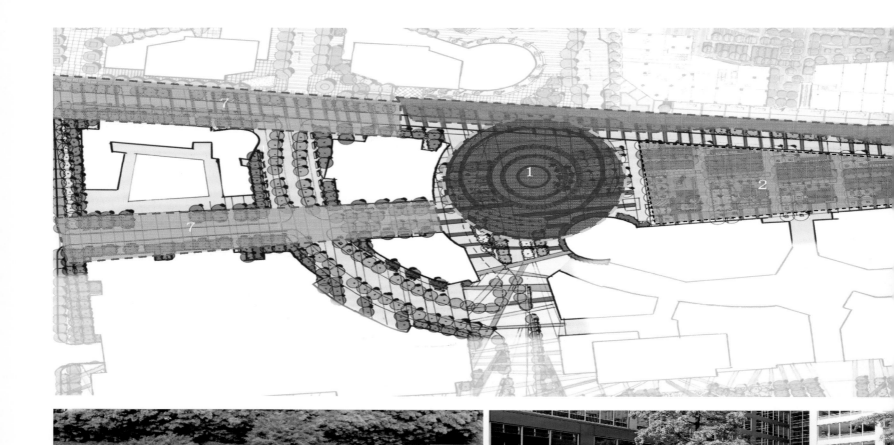

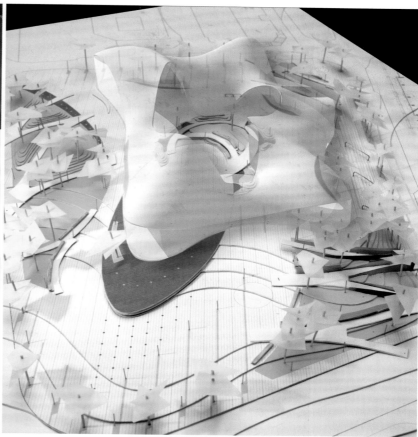

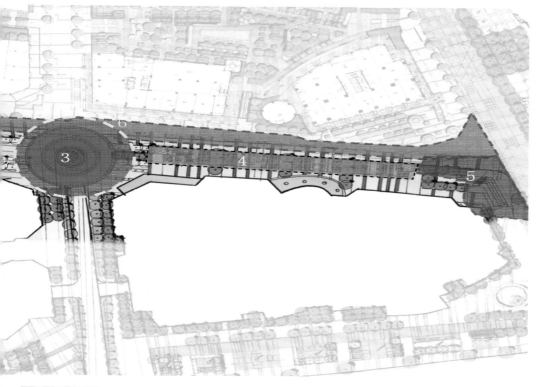

### Overview 项目概况

Taskin City Locates in the interjection between Xiangfu Road and Shaoshan Road, north is adjacent to Yinxin Road, east is near with Shaoshannan Road and south is close to Xiangfu Road. This project covers an area of 374,000 m², a comprehensive architecture complex with multiple activities like commercial, retail, general merchandise, hotel, office, catering, theme commercial street, fashion shopping, entertainment, apartment, children's space, education etc. into one.

湖南长沙德思勤城市广场位于湘府路与韶山路交汇处，北临迎新路，东临韶山南路，南临湘府路，本项目占地面积374 000 m²，是一个集商业、零售、百货、酒店、办公、餐饮、主题商业街、时尚淘宝、娱乐、公寓住宅、儿童天地、教育等多种业态为一体的综合性建筑群。

| | | | | |
|---|---|---|---|---|
| 1 | 舞台区景观空间 | | 5 | 开场入口景观空间 |
| 2 | 开敞互动景观空间 | | 6 | 景观走廊空间 |
| 3 | 重要水景节点空间 | | 7 | 轴线延伸景观空间 |
| 4 | 休憩互动景观空间 | | | |

## ▶ Landscape Feature  景观特色

Rising Star, prosperity of time and prospects capital are the core in the theme design concept of Taskin City. They separately use composite activity forms like hotel, food court, financial centre and fashion catering etc. to construct theme concept, making them to be the feature and highlight for theme creation.

The stage design turns to be the carrier of enriching citizen life and broadcasting Changsha culture and to be the landscape focus in terms of form and function, as well as building high quality landscape access space, stop space and viewing space which combines axis commercial interface. That makes the axis gathering stage, commercial street to be a vital landscape district and provides significant commercial value and social value.

The design insists in the concept that given consideration to natural ecology and entertainment shopping. Through analysis, the designer is creating a rest space in the street's commercial district design, and is creating a multi-functional garden park suitable for passing, stop and rest use while combining plant groups and meeting the requirement for residential and commercial use.

德思勤城市广场在主题概念设计中，分别以明日之星、时代之华、展望之都为核心，并以酒店、美食街、金融中心、潮流餐饮等复合业态形式来构建主题理念，使之成为主题营造的特色及亮点。

通过对舞台的设计，使其成为传播长沙文化、丰富市民生活内容的载体，并在形式和功能上成为景观聚焦点。并结合轴线商业界面营造高品质的景观通道空间、停留空间、观赏空间，使这条具有舞台、商业景观街两大功能的轴线成为重要的景观区域并产生大的商业价值与社会价值。

设计中坚持自然生态与休闲趣味购物兼顾的设计理念，通过分析，在沿街商业区域设计停留休憩的功能空间，并结合绿植群落营造适宜行走、停留、休憩的多功能园林式空间，满足住商两用的功能需求。

## COMMERCIAL LANDSCAPE
## 商业景观

## MODERN STYLE
## 现代风格

### KEY WORDS 关键词

**LANDSCAPE EFFECT**
造景效果

**LAMP BELTS**
发光灯带

**LANDSCAPE NODE**
景观节点

Location: Suzhou Industrial Park, Jiangsu
Developer: Harmony Group
Landscape Design: Shenzhen DongDa Landscape Design Co., Ltd.
　　　　　　　　SWA Group
Land Area: 80,000m²

项目地点：江苏省苏州市
开 发 商：苏州圆融集团
景观设计：深圳市东大景观设计有限公司
　　　　　美国SWA景观设计事务所
占地面积：80 000 m²

# Suzhou Harmony Times Square
# 苏州工业园圆融时代广场

### FEATURES 项目亮点

With a landscape street, three typical corridors, five joints and nine spaces, it creates a comprehensive square space.

项目以一个总体街道景观框架、三个具有特色的廊道、五个连接节点以及九个场所为结构基础，打造了一个综合性的广场空间。

## ▶ Design Theme 设计主题

The theme of the design is to "conform to the trend of the times". The square is composed of a landscape street, three typical corridors, five joints and nine spaces. In the landscape design, all the details are carefully considered and exquisitely designed.

　　项目设计主题是"时代的脉动和时代的潮流"。广场以一个总体街道景观框架、三个具有特色的廊道、五个连接节点、九个场所为结构基础，综合各节点细节的设计思路充分体现设计精髓。

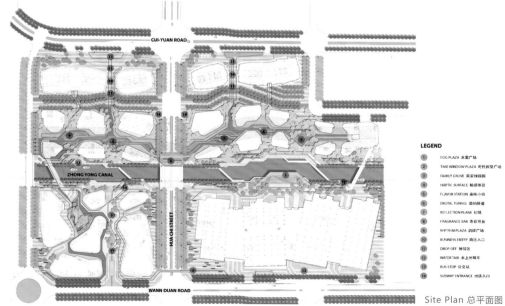

Site Plan 总平面图

## ▶ Landscape Features  景观特色

During the construction, the designers have to find solutions to many problems, for example, the technology for energy bands installation, the structure and eco design of the revetments, the second-design of the landscape bridge, and replacement of the entrance materials, the second-design of the spray square, the drainage of the plant area, etc.

Energy bands can be found everywhere on the square ground. Decorated with lamps, they create special lighting effect. To achieve this effect, it must solve two problems.

First, the energy bands should meet the firefighting truck load requirement. Second, the outdoor lamps should be sheltered from rain. With these restrictions, the designers give the solution to use glass bricks and LED lamps. At the same time, the existing revetments are redesigned to keep the original design and enhance the ecological quality of this area. In terms of the bridge construction, the architects renew the steel structure design to make the slope gentle for easy construction. And at the same time, it well keep the landscape view of the bridge.

在项目实施中，设计师攻克众多难题，诸如能量带的材料实施技术，河道驳岸的结构及生态实施，跨河景观桥的二次结构设计，轨道交通出入口的材料更替设计，水雾广场的二次方案实施以及植物的排水问题等。

能量带遍布广场地面，由发光灯带形成线形光线效果。在实施过程中需要解决以下两个问题：第一，能量带必须满足消防车的荷载；第二，室外地面灯带必须免受雨水侵蚀，方便日后维护。在这两个条件的限制下，设计师提出采用玻璃砖及 LED 灯相结合的方式，较大程度地解决了能量带实施难、养护难的问题。同样，在河道驳岸实施中，设计师在原有市政河道的驳岸结构基础上，进行二次设计，保留了原有设计初衷，并增强了河道驳岸的生态性。在跨河景观桥的实施中，鉴于 SWA 设计的桥身系多方向扭曲体，施工难度大，行走不便，遂进行了二次钢构设计，增加桥身平缓度，便于施工，亦保证了整体桥身景观效果。

# COMMERCIAL LANDSCAPE
商业景观

## NEOCLASSICAL STYLE
新古典主义风格

### KEY WORDS 关键词

**EXQUISITE AND RICH**
细腻丰富

**LANDSCAPE CHARACTERISTIC**
景观特色

**LANDSCAPE NODE**
景观节点

---

Location: Chongming, Shanghai
Developer: Shanghai Xinchong Construction and Development Co., Ltd.
Landscape Design: Shanghai WEME Landscape Engineering Co., Ltd.
Landscape Area: 24,154 m²
Total Floor Area: 52,412.98 m²
Total Land Area: 33,227 m²

项目地点：上海市崇明县
建设单位：上海新崇建设发展有限公司
景观设计：上海唯美景观设计工程有限公司
景观面积：24 154 m²
总建筑面积：52 412.98 m²
总用地面积：33 227 m²

# Chongming Yulin Business Square
崇明育麟商务广场

## FEATURES 项目亮点

The design carries on the architectural style, using classical architectural symbols and modern materials to create a delicate, leisure and comfortable environment with strong artistic breath.

设计手法上延续建筑的风格，利用建筑的古典符号和现代材料，整体上创造精致、闲适又有浓厚艺术气息的环境。

### ▶ Overview  项目概况

The project is located in the intersection of Linqiao Road and Gulangyu Road in Chengqiao Town, Chongming County, including hotel landscape and an urban public green land. This land has open urban space, greatly enhancing the city image of Chengqiao Town and even the Chongming Island.

项目位于崇明县城桥镇育麟桥路和鼓浪屿路交叉口，包含酒店景观和一块城市公共绿地。此地城市空间开阔，具有提升城桥镇乃至崇明岛的城市形象的作用。

### ▶ Design Concept  设计理念

Through the detailed demonstration of the landscape and the overall configuration, the project highlights the luxury and solidness of Art Deco architecture, exquisite and rich detailed decoration and the elegance and steady of theme landscape; it also blends in the delicate, introverted and leisure temperament of the Shanghai regional culture to implement the perfect combination of architectural landscape art and the city life.

项目通过景观的细节描绘及整形配置凸显 Art Deco 建筑的华贵坚实、细部装饰的细腻丰富、主题景观的典雅稳重，并融汇了上海海派文化的精致、内敛、闲适的气质，实现了建筑景观艺术与都市生活的完美结合。

Site Plan 总平面图

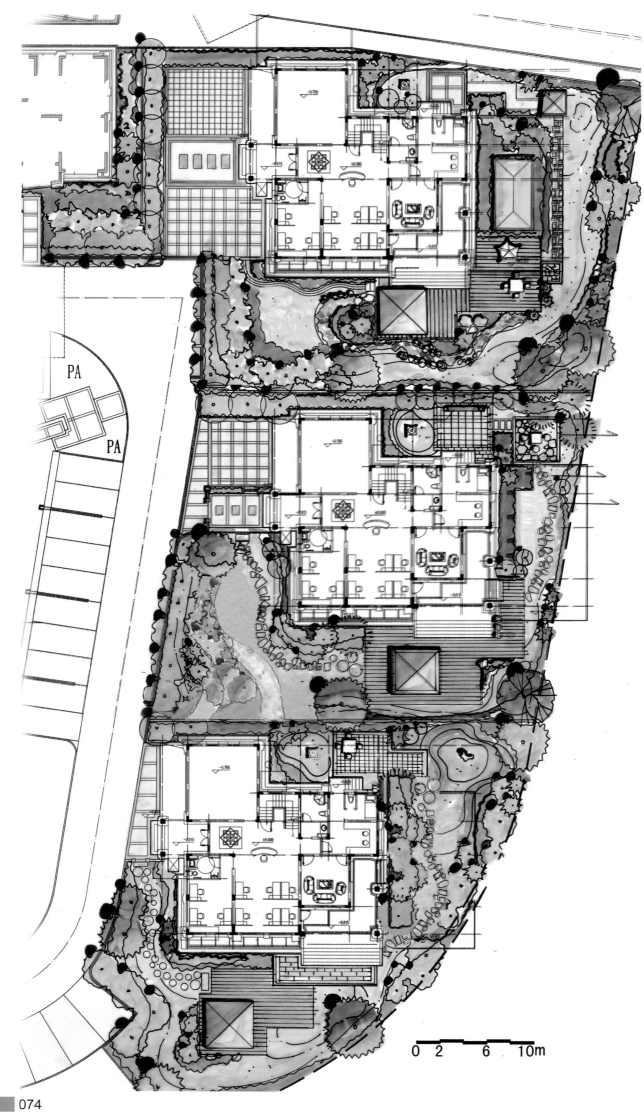

## Landscape Features 景观特色

Different functional areas are independent in the overall layout; under the precondition of reasonable and efficient transportation, it realizes the mutual penetration of the regionally visual landscape and the functional facilities. Due to the relatively insufficient external landscape in the office area, it creates rich interior landscape; the hotel leads in the urban green landscape and pluses its own landscape system to bring the consumers strong visual enjoyment. Both the architecture and landscape bring people a kind of shock on the soul.

The landscape features are embodied in the perfect combination of classical and modern, and the classical symbols and modern materials create a kind of brand-new visual impact. The translucent materials used in the office and hotel avoid the interference of sightline and give the building a hazy beauty, making the building itself become an art work. In the daytime, the elegant and fashionable buildings look like the static sculptures, standing on the street corner. When the evening lights are lit, these resplendent and magnificent buildings add a dash of bright color for the urban landscape.

　　整体布局上不同功能区域各自独立,在合理高效的交通前提下,满足区域视觉景观和功能使用上的相互融会贯通。办公区域由于外部景观相对不足,因此创造了丰富的内部景观;酒店部分借用城市绿地景观,加上基地内部自身的景观体系,均带给消费的人群强烈的视觉享受。建筑和景观共同带给道路上的行人一种心灵的震撼。

　　景观特色体现在古典与现代的完美结合,古典的符号与现代的材料缔造出一种全新的视觉冲击力。办公和酒店部分通过对半透明材质的利用,避免了视线干扰,同时使建筑产生朦胧的美感,使得建筑本身成为一件艺术品。白天,典雅时尚的建筑犹如一座座静态的雕塑,矗立在街角。夜晚华灯初上,金碧辉煌的建筑形体为城市景观增添了一抹亮色。

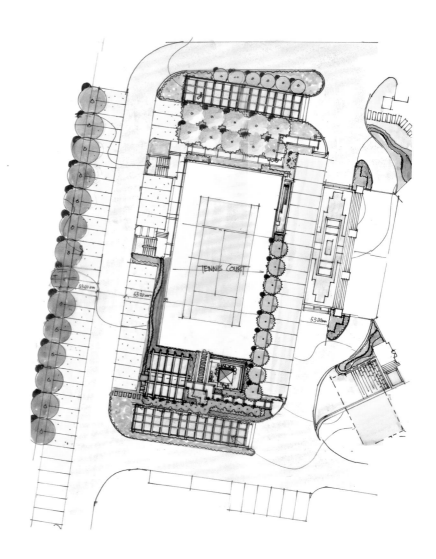

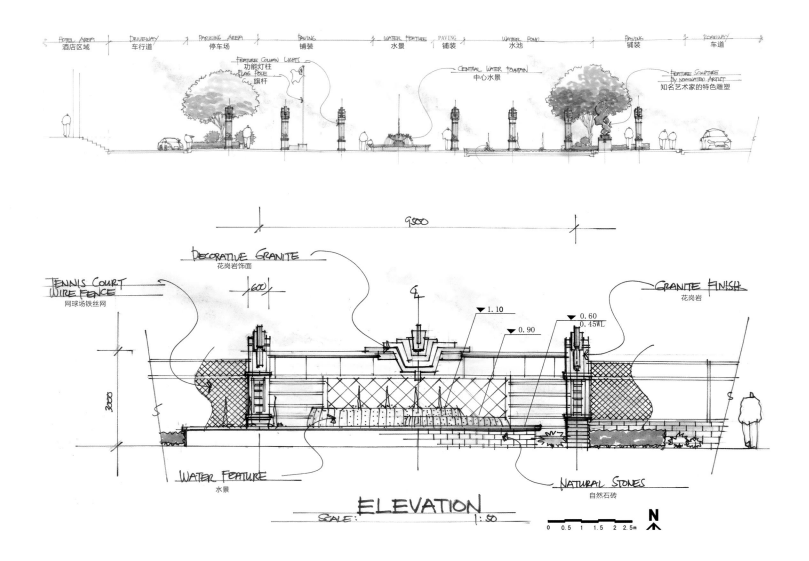

# Tourism Landscape
# 旅游度假区景观

**Cultural Environment**
人文环境

**Recreation and Entertainment**
休闲娱乐

**Eco Space**
生态空间

# TOURISM LANDSCAPE 旅游度假区景观

## GARDEN VILLAGE STYLE 花园式农庄风格

### KEY WORDS 关键词

- LANDSCAPE NODE 景观节点
- EXPERIENTIAL LANDSCAPE 体验式景观
- ECOLOGICAL ENVIRONMENT 生态环境

Location: Taizhou, Zhejiang
Planning and Design: Hangzhou Modern Environmental Art Co., Ltd.

项目地点：浙江省台州市
规划设计：杭州现代环境艺术实业有限公司

# The Overall Planning of Taizhou Xianju Tea Valley Resort

## 台州仙居茶溪谷度假区总体规划

### FEATURES 项目亮点

The project is based on the superior ecological environment with peculiar farm experience resort features and obvious regional characteristics.

项目基于优越的生态环境，以特有的农场体验式度假为特色，地域特质明显。

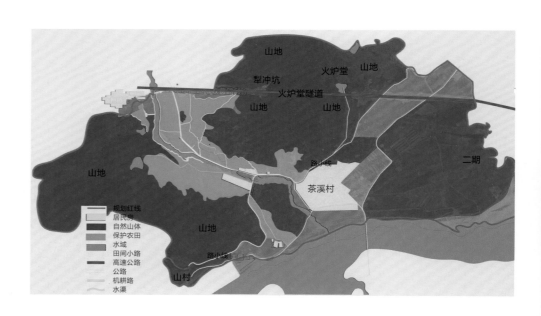

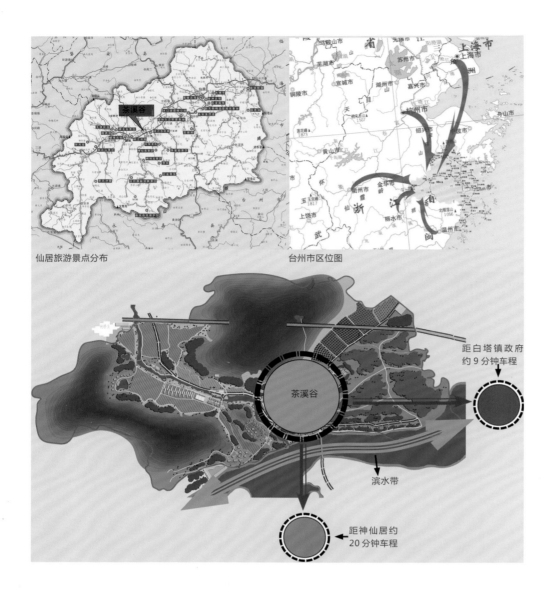

## Overview 项目概况

The project is located in Baita Town, Xianju County- the middle part of Xianju County. Tea Valley Farm is not only a three-high (high yield, high quality, high benefit) demonstration base of agricultural scientific research and production, but also a garden-village-type integrated resort with facilities of leisure, vacation, tourism, entertainment, sports, shopping, meeting, scientific research and popular science, which is a good place for leisure, tourism, vacation, entertainment and the ideal base for team developing training, conference and group activities.

仙居茶溪谷度假农场位于浙江省仙居县白塔镇，地处仙居县中部。茶溪谷度假农场既是三高（高产量、高质量、高效益）农业生产科研示范基地，又是一个具有花园式农庄风格的集休闲、度假、观光、娱乐、运动、购物、会议、科研、科普于一体的综合度假胜地，是人们休闲、旅游、度假、娱乐的好去处，也是团队拓展训练、会议、集体活动的理想基地。

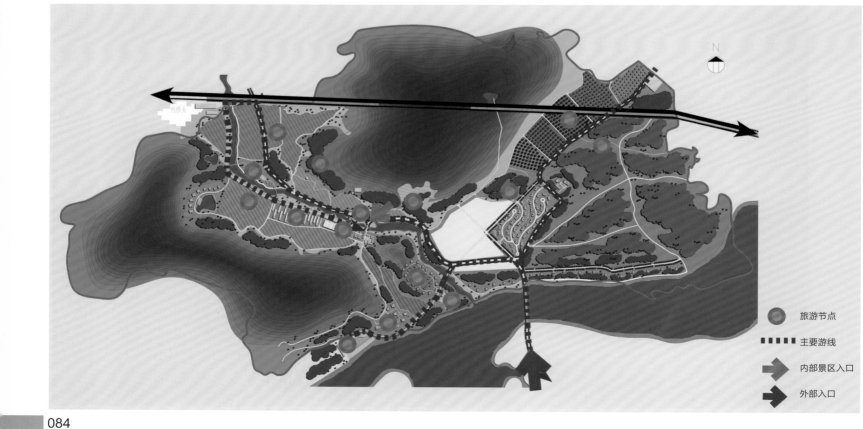

## ▶ Design Concept  设计理念

This project is based on the excellent ecological environment, beautiful fairy local residences, unique farm-experience-type resort, cultural & creative agriculture to create large-scale resort-type farm with rich regional characteristics of ecological agriculture, leisure healthy living and recreation experience.

本项目以良好的生态环境为基础，以优美的神仙居"仙人居住的地方"为依托，以特有的农场体验式度假为特色，以文化创意农业为亮点，全力打造富有地域特色的集生态农业、度假、休闲、养生和游乐体验为一体的大型度假农场。

## ▶ Ecological Measures  生态措施

The project sticks strictly to the management measures for ecological environment by strengthening the solid pollution noise processing, water environmental protection measures, noise & air pollution treatment, landscape protection and green building to achieve the aim of environmental conservation & maintenance purposes and minimize damages to the ecological environment. The forest planning, waterscape construction and landscape creating enhance the environmental benefits in that region.

项目建设实施严格的生态环境管理措施，通过加强固体污染物的处理、水体环境保护措施、噪声的处理、对大气污染物的处理、景观特色保护措施和绿色建筑，达到环境保育维护的目的，最大限度地减少对生态环境的损害，以林相改造、水景疏建、景观营造等措施增强区域的环境效益。

# TOURISM LANDSCAPE
## 旅游度假区景观

## MODERN STYLE
## 现代风格

### KEY WORDS 关键词

**ECOLOGICAL NATURE**
生态自然

**HUMANITY LANDSCAPE**
人文景观

**LANDSCAPE NODE**
景观节点

Location: Jiexiu, Shanxi
Landscape Design: Hangzhou Modern Environmental Art Co., Ltd.
Total Land Area: 647 005.21 m²
Total Floor Area: 135 200.95 m²

项目地点：山西省介休市
景观设计：杭州现代环境艺术实业有限公司
总用地面积：647 005.21 m²
总建筑面积：135 200.95 m²

# Shanxi Zhangbi Castle SPA Resort
## 山西张壁古堡温泉养生度假区

### FEATURES 项目亮点

The project takes healthy resort, military culture and constellation culture as the subject to highlight the characteristics of health, leisure and experience.

项目以养生度假、军事文化、星宿文化为主题，突出健康、休闲、体验的特色。

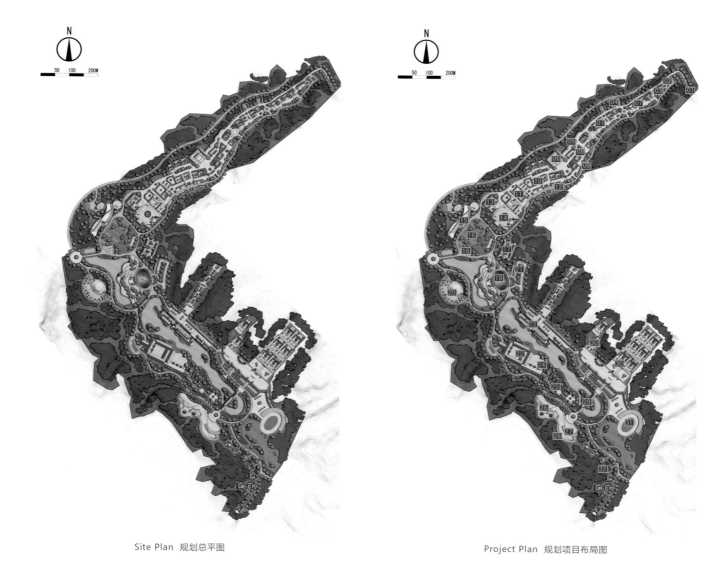

Site Plan 规划总平图

Project Plan 规划项目布局图

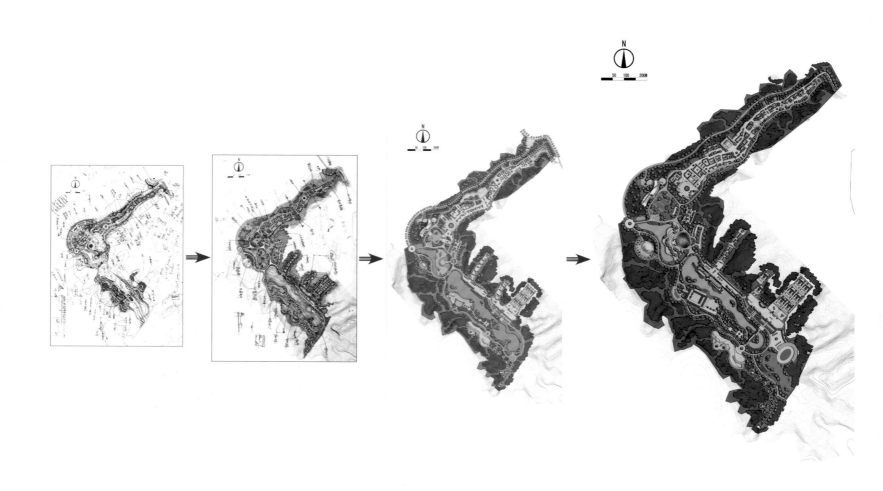

## Overview 项目概况

The project takes healthy preserving resort, military culture and constellation culture as the subjects to make a bodybuilding and healthy SPA town. The project selection and service supporting facilities show that the planning area provides the visitors a leisure and healthy life and highlights the health preserving SPA and the culture of the town.

The project takes advantage of the natural layout of terrain changes in the form of ladder like backward extension. The environment is created on the basis of the original terraced fields, wetlands, clear spring and other natural elements between the plateau and canyon, and through this original ecological space layout, the pastoral interest and pure vocational springs bring out the best in each other.

项目以养生度假、军事文化、星宿文化为主题，打造康体养生型温泉小镇。从项目选择、服务配套等凸显规划区为游客提供的闲适健康生活，突出温泉养生、小镇文化。

项目的建设多采用依托地形变化形成的阶梯式后退延展的自然布局。环境的营造借助高原、峡谷之间原有的梯田、湿地、清泉等原生态自然元素，通过原生态空间布局，田园野趣与纯正度假温泉相得益彰。

## Design Concept 设计理念

Relying on the Zhangbi castle culture, the project highlights the natural resources of loess grand sight and becomes "a tourism distribution center in Jinzhong". Combined with the design idea and the characteristics of local nature & humanity landscape resource, it takes family, unit groups, government reception, Self-driving tourists and the elderly health care as the main market; based on the typical loess plateau terraced landscape and the strong castle culture, it builds a hot spring resort town with rich culture contents of hot spring health preservation, business meetings, bodybuilding and cultural experience, which also has safe, comfortable and beautiful environment, perfect service facilities, business management and flexible management mechanism.

设计旨在依托张壁古堡文化，突出"黄土大观"自然资源，打造成为"晋中地区旅游集散中心"。结合方案构思与当地自然、人文景观资源特点，以家庭、单位团体、政府接待、自驾车游客、中老年人养生保健为主要市场；以典型的黄土高原梯田景观风光为特色，以厚重的古堡文化为底蕴，打造具有安全舒适、环境优美、服务设施完善、经营管理机制灵活，集温泉游乐养生、商务会议、康体健身、文化体验为一体的文化气息浓郁的温泉休闲度假小镇。

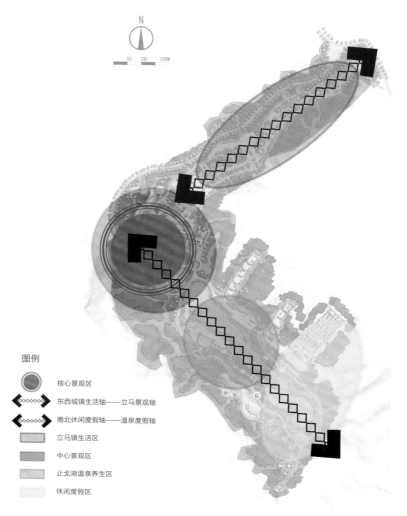

Space Layout 空间布局图

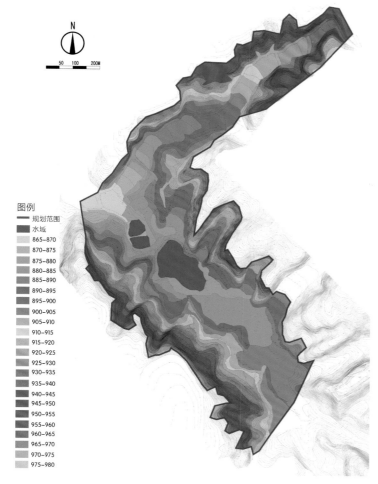

Altitude Analysis 基地高程分析图

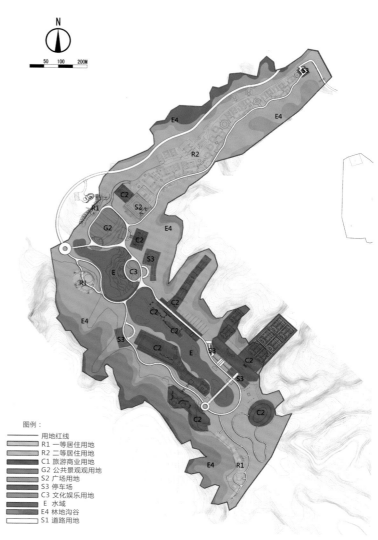

Functions Arrangement 功能分区图

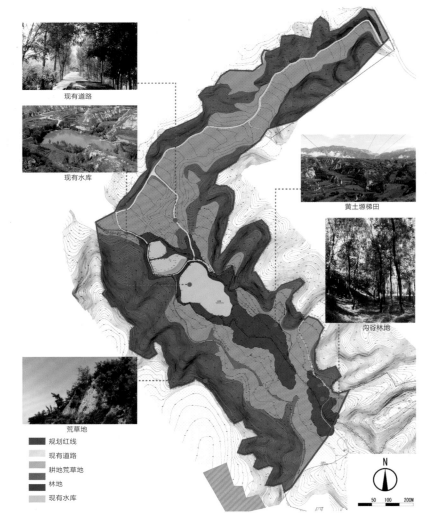

Resource Condition Analysis 资源现状分析图

### Cultural Features 文化特色

The project fully displays the local culture and the active culture and especially highlights the castle culture, local military culture and health preserving SPA culture. The concept of "activity" is put into the resort area and the idea of "living" is drawn into the landscape design. It builds a landscape SPA town to make tourism activities be part of the daily life, which brings new significance, objective, methods and means to tourism development to change the past pure planning concept of tourist destination.

做足地方文化、活化文化，突出古堡文化、军事文化等本土文化及温泉本身的养生文化。将"活动"引入度假区内，把景观植入"活"的概念。规划一处可以参与的景观温泉小镇，使旅游活动变成日常生活的有机组成部分，赋予了新的旅游意义、目的、方式和手段，改变了以往纯粹旅游目的地的景区规划概念。

### Environment Features 环境特色

On one hand, the project pays attention to the construction of outdoor plateau terraced landscape and leisure space, forming a "hot spring town" with laid-back vocation atmosphere; on the other hand, with the utilization of unique military culture, constellation culture and spring culture to make profound meaningful landscape environment.

一方面注重户外高原梯田景观、休闲空间的营造，形成"温泉小镇"的慵懒闲适的度假氛围；另一方面，利用古堡独特的军事、星宿文化结合温泉，营造具有深厚内涵的景观环境。

# TOURISM LANDSCAPE
旅游度假区景观

## MODERN STYLE
现代风格

### KEY WORDS 关键词

**LANDSCAPE NODE**
景观节点

**PLANT LANDSCAPING**
植物造景

**ECOLOGICAL GREENBELT**
生态绿带

Location: Kashi, Xinjiang Uygur Autonomous Region
Developer: Kashi North Real Estate Development Co., Ltd.
Landscape Design: Guangzhou Taihe Landscape Design Co., Ltd.
Area: about 2,000,000 m²

项目地点：新疆维吾尔自治区喀什市
开 发 商：喀什北湖房地产开发有限公司
景观设计：广州市太合景观设计公司
面　　积：约 2 000 000 m²

# Kashi North Lake Ecological Park Landscape Planning and Design

喀什北湖生态公园景观规划概念设计

## FEATURES 项目亮点

The project has water as the main body, the mountains as the foil and the green resources as the scenery source.

以水为主体，以山为衬托，以绿为景源。

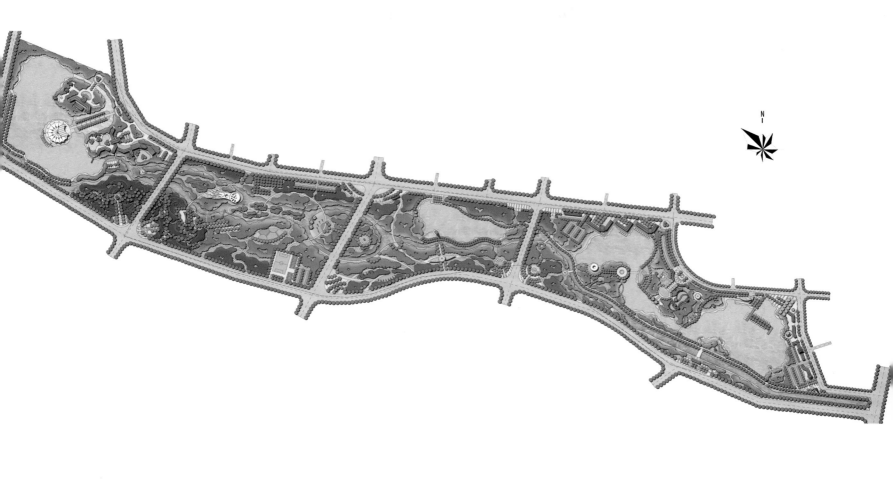

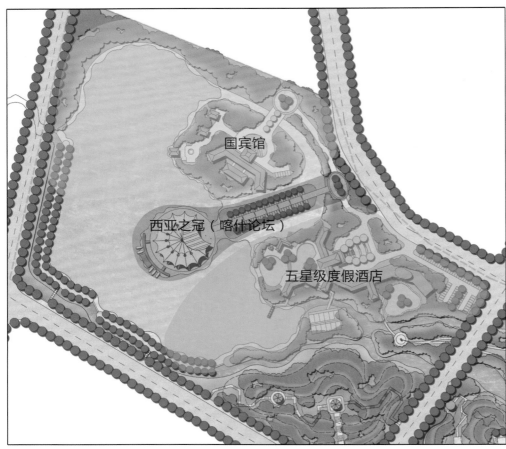
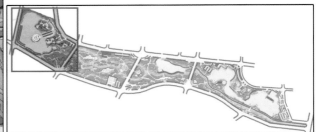

International Tourist Resort Plan 国际旅游度假区分区图

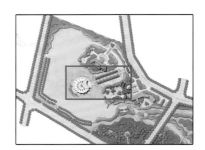

1. 星光大道
2. 停车位
3. 景观灯柱
4. 喀什论坛
5. 亲水木平台
6. 渔人码头

Plan 分区设计图

### Overview 项目概况

North Lake Park is located in the northwest of Kashi City, the upstream of Tuman River, with an area about 2,000,000 m². The park is built on the basis of Tuman River; near the river bank there are natural fish ponds, plants and animal activities around, and the surrounding farmland grow well which is also suitable for planting.

北湖公园位于新疆喀什市西北部，吐曼河流域的上游区域，面积约 2 000 000 m²。公园以吐曼河水系为基础，河堤岸附近已自然形成鱼塘，周围有植物生长与动物活动，周边农田生长状况良好，适宜种植。

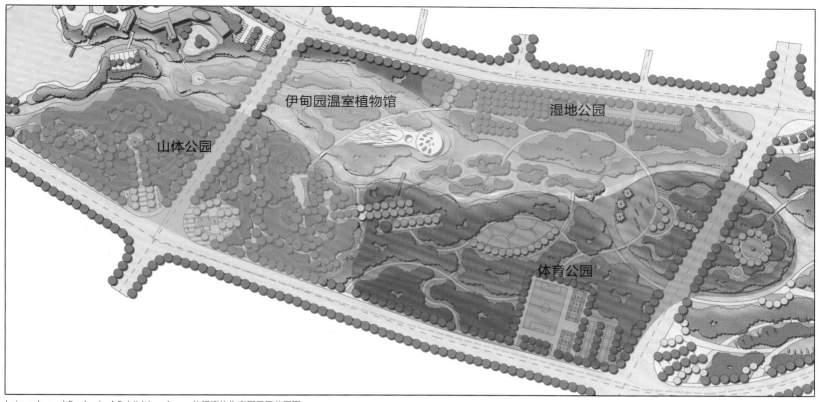

Leisurely and Ecological Exhibition Area 休闲康体生态展示区分区图

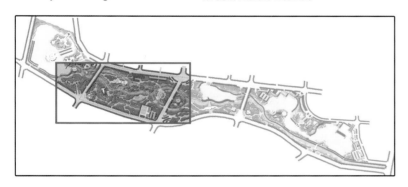

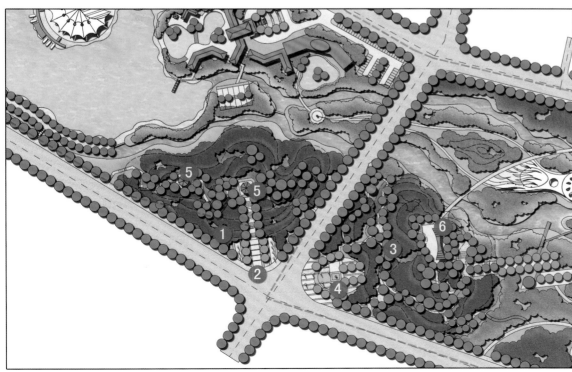

Plan 分区设计图

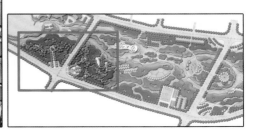

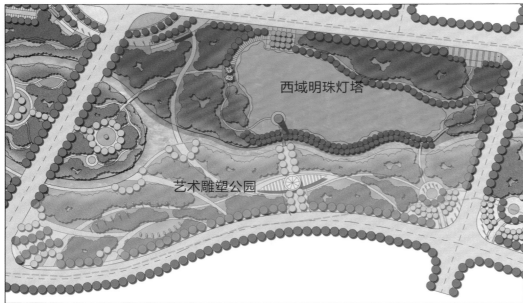

Cultural and Historical Area 历史文化区分区图

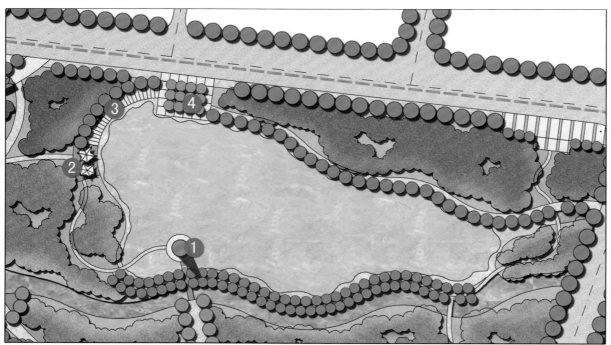
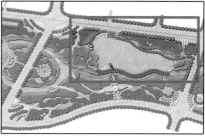

Plan 分区设计图

## Design Concept 设计理念

In the development of modern city, the catalysis of water is obvious indeed. According to the development trend of modern waterfront park tourist area, the definition of the district by the overall planning of Kashi City, the geographical characteristics of Tuman River and the surrounding green belt, geographical factors, historical context and the surrounding environmental conditions and other elements, the project takes the strategies of industry planning, landscape recreation, ecological restoration, function promotion and keeps Tuman River ecological waterfront greenbelt as the supporting point to preserve the existing waterfront ecological green belt, refine and sublimate the Xinjiang Kashi historical art culture & folk customs and combine with the industries of leisure, tourism and vacation, thus building the North Lake Park District into an urban boutique scenery district including tourism and sightseeing, commercial exhibition, recreation, community, learning and leisure, which also adheres to the times development requirements of regionalization, folk custom keeping, ecological, industrialization.

现代城市发展中，水体的触媒作用显著。根据现代滨水公园旅游区的发展趋势及喀什市总体规划对本区的定位，综合考虑吐曼河及周边绿带的地理特点、地缘优势、历史文脉及周边环境条件等多方面的因素，项目将通过产业策划、景观重塑、生态修复、功能提升等策略，以吐曼河生态滨水绿带为载体，一方面尽量保留现有滨河生态绿带，同时对新疆喀什历史文化艺术、民族风情进行提炼与升华，并且与休闲、旅游、度假相结合，将北湖公园区建设成集旅游观光、商务会展、游乐、参与、求知、休闲于一体，适应时代发展要求的地域化、民俗化、生态化、产业化的都市精品风景区。

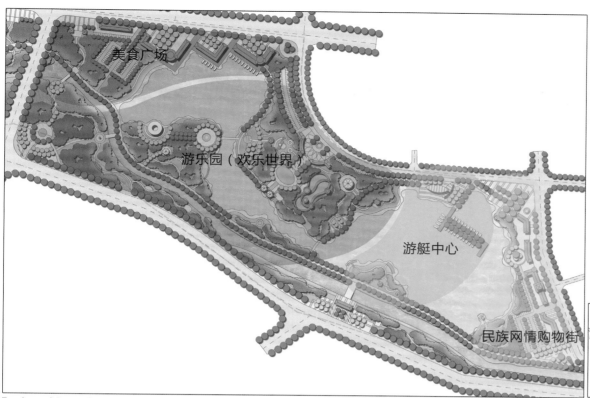

Tourist and Entertainment Area 旅游包含娱乐区分区图

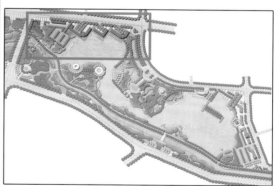

Plan 分区设计图

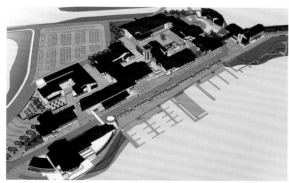

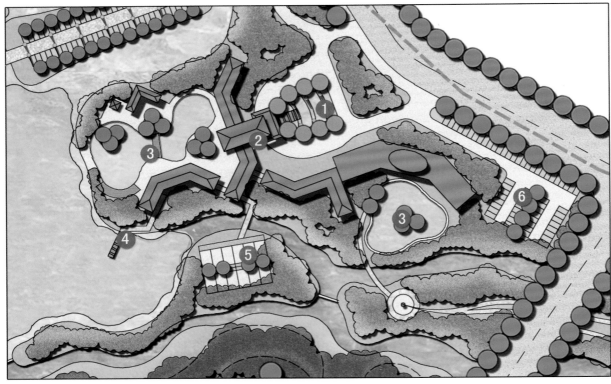

Plan 分区设计图

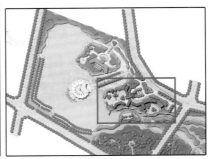

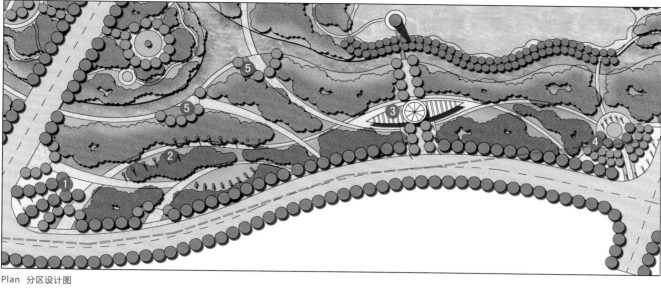

Plan 分区设计图

### ▶ Ecological Landscape System 景观生态系统

Lakes and Rivers: The park is located in the upstream of Tuman River, having Tuman River system as the main water source. It carries on the pollution control and ecological construction at the same time, implementing the river dredging and solid waste clean-up project through artificial river sediment and the ecological purification system built by the Wetland Park, which brings the Tuman River "clear water and green shore ". The three new huge lakes, one is romantic and quiet, one is quiet and fresh and the third one is vibrant and dynamic, so that the whole park keeps the water as the main body, the mountains as the foil and the green resources as the scenery source, together with the square, buildings and other landscape elements to perfectly combine the fashionable & dynamic city life with the quiet natural environment.

Greening System: The plants are mainly native species and the arbors, shrubs and ground flora are matched perfectly together. According to the landscape function and use function it creates rich or simple plant levels. In the planning and design, evergreen and deciduous plants are combined together, fast-growing, fast-less-growing and slow-growing trees are planted together, and according to the plant morphology, color, smell and texture, the plants will be matched in reasonable and art way, which creates a stable but four-season different plant landscape.

Square: Plants growing in the square are mainly tall and beautiful trees and low shrubs, together with small trees to create simple plant levels, thus bringing a beautiful, open, leisure and comfortable garden space with cool summer and warm winter.

Road: Road Greening should bear the functions of overshadowing, dust-cleaning, wind resistance, anti toxic gas in addition to beautifying the city. In the selection of plants, it mainly chooses large deciduous trees, or deciduous or evergreen small trees accompanied by the bright color low shrubs, flowers to create spacious, bright and comfortable road landscape.

Open Space: The adoption of sparse forest grassland and open lawn and the rich changing colors in the space boundary bring visitors sufficient moving space. The sparse forest and shadows adorned with some low blossoms or foliage coverings with rich colors in four different seasons can bring out the characteristics of peaceful, active and elegant atmosphere of the park.

Ecological Control Zone: It keeps to the principles of ecological priority and species diversity. Through abundant plant species, it creates a good living environment for a variety of microorganisms and small animals, forming a relatively stable wetland ecosystem.

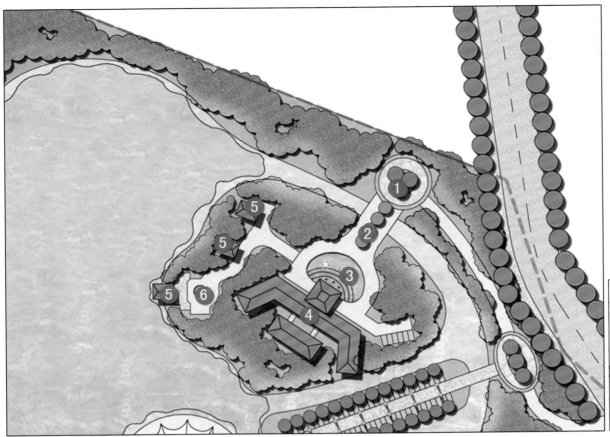
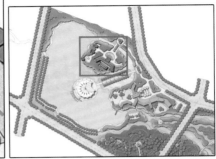

Plan 分区设计图

　　湖泊、河流：公园位于吐曼河流域的上游，以吐曼河水系为主要水源基础，采取污染治理与生态建设相结合的方式，实施河道清淤、固体垃圾清理工程，通过人工河道沉沙及湿地公园自建立的生态净化系统，让吐曼河"水清、岸绿"。重新规划设计的三个大湖面，一个浪漫幽静，一个静谧清新，一个活力动感，使整个公园以水为主体，以山为衬托，以绿为景源，与广场、建筑等景观元素融合一体，将时尚动感的城市生活与宁静的自然环境完美结合。

　　绿化系统：以乡土植物为主，将乔灌木、地被巧妙地配搭在一起，根据景观功能和使用功能的需要营造或丰富或简洁的植物层次。设计中注意常绿植物与落叶植物相结合，速生与中生、慢生树相结合，并根据植物的形态、色彩、气味、质地等进行合理、艺术的配搭，营造稳定而四季各异、四季皆景的植物景观。

　　广场：广场的植物以高大优美的乔木和低矮的花灌木为主，辅以中小乔木，通过构造简洁的植物层次，营造一个优美、开阔、休闲、舒适、冬暖夏凉的园林空间。

　　道路：道路绿化需在满足美化的同时起到遮阴、滞尘、抗风、抗有毒气体等作用。在植物的选择上，以落叶大乔木为主，以落叶或常绿的中小乔木适当配以色彩明丽的低矮灌木、花草，营造一个宽广、明亮舒适的道路景观。

　　开放空间：多运用疏林草地以及开敞的大草坪形式，在空间边界限定上实现丰富多彩的变化，给予游人充分的活动空间，在疏林或林荫下，配以成片低矮的不同时令开花或观叶的植被以丰富四季色彩，烘托公园或宁静、或活泼、或大气的特点。

　　生态控制带：遵循生态优先原则、物种多样性原则，通过丰富的植物种植为各种微生物、小动物创造良好生境，形成相对稳定的湿地生态系统。

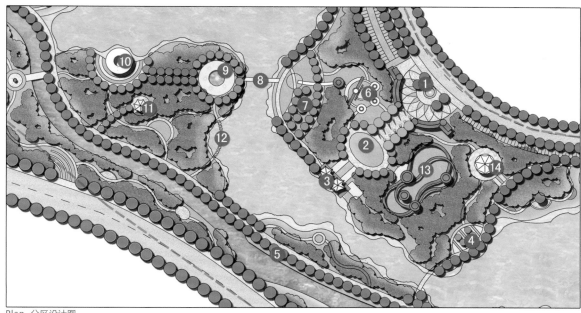
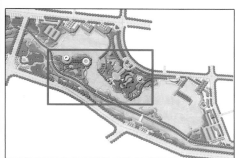

Plan 分区设计图

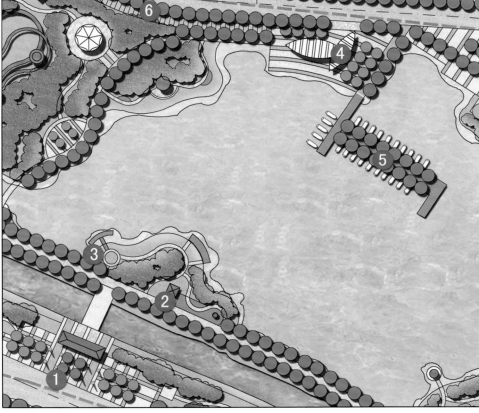
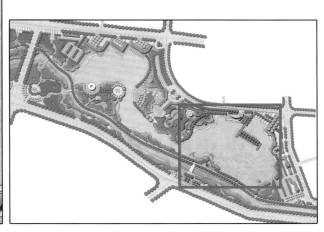

Plan 分区设计图

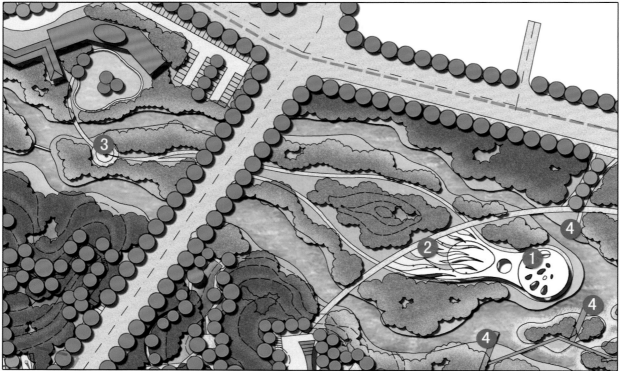

Plan 分区设计图

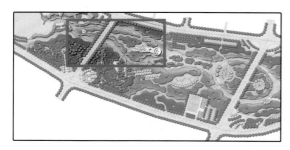

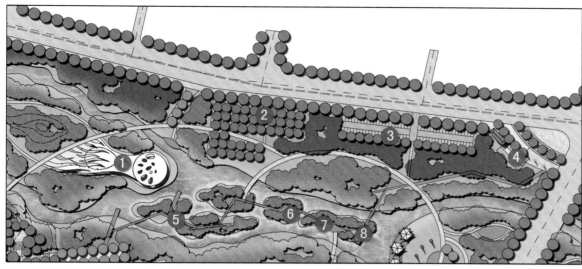

Plan 分区设计图

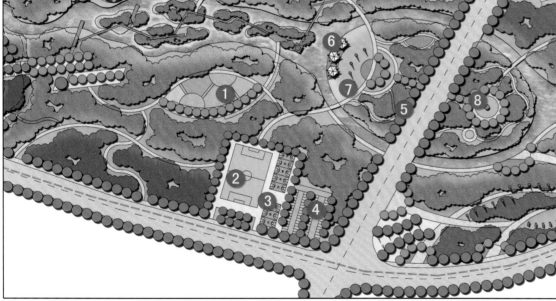
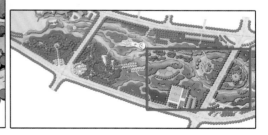

Plan 分区设计图

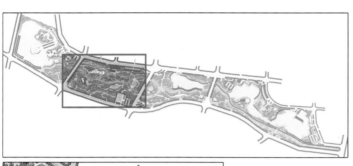

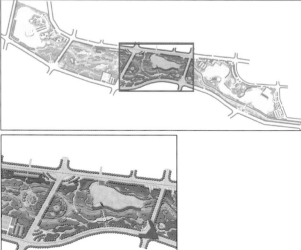

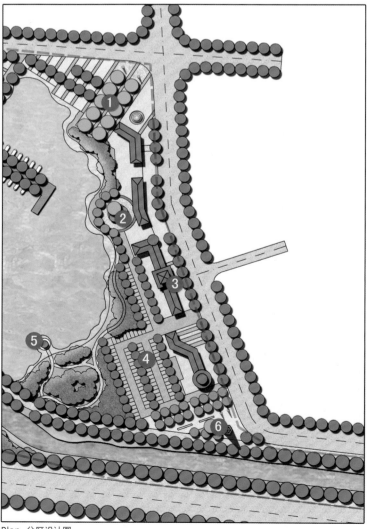
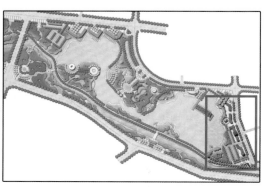

Plan 分区设计图

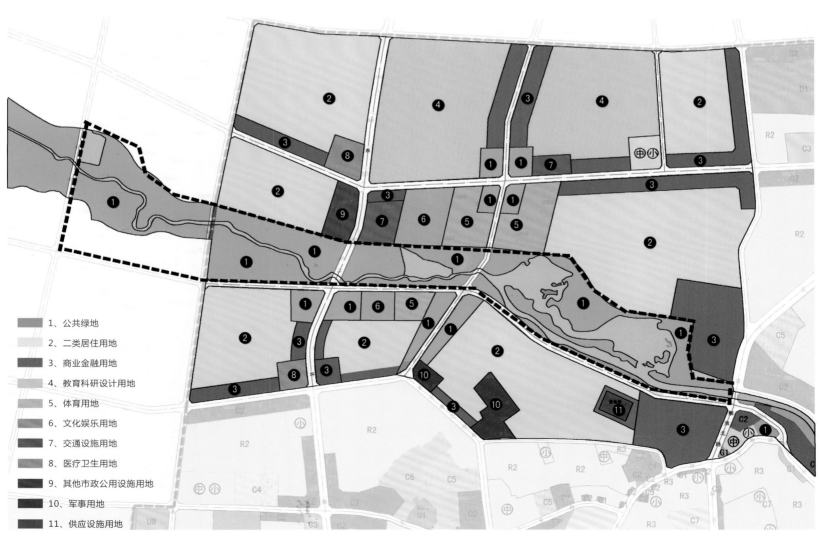

- 1、公共绿地
- 2、二类居住用地
- 3、商业金融用地
- 4、教育科研设计用地
- 5、体育用地
- 6、文化娱乐用地
- 7、交通设施用地
- 8、医疗卫生用地
- 9、其他市政公用设施用地
- 10、军事用地
- 11、供应设施用地

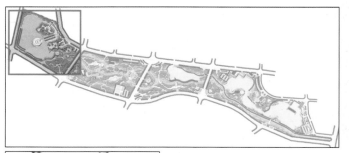

# TOURISM LANDSCAPE
旅游度假区景观

## MODERN STYLE
现代风格

### KEY WORDS 关键词

- ECOLOGICAL SPACE 生态空间
- LANDSCAPE NODE 景观节点
- MOUNTAIN LANDSCAPE 山体景观

Location: Dalian, Liaoning
Developer: Dalian Yida group / Dalian Software Park Co., Ltd.
Landscape Design: Pasno (Asia) Landscape Design

项目地点：辽宁省大连市
开 发 商：大连亿达集团 / 大连软件园股份有限公司
景观设计：普梵思洛（亚洲）景观规划设计事务所

# Dalian Software Park Donggou Mountain Park

大连软件园东沟山体公园

## FEATURES 项目亮点

Through the optimization design on the mountain, the project creates an international Mountain Ecological Park with the functions of leisure, culture and fitness as a whole.

通过对山体的优化设计，打造出集休闲、文化、健身为一体的国际化山体生态公园。

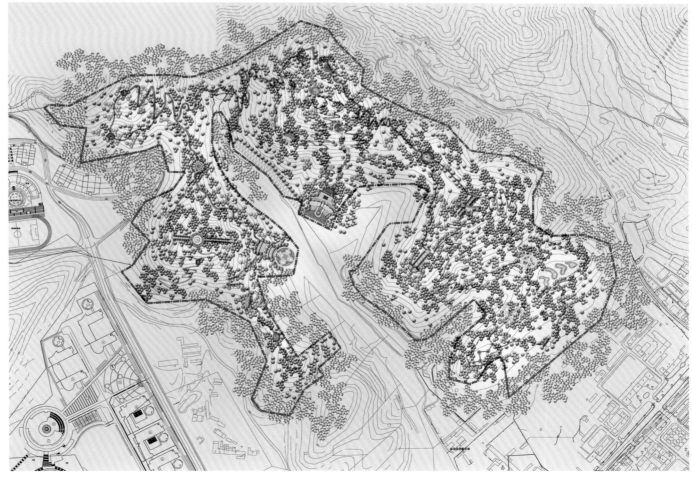

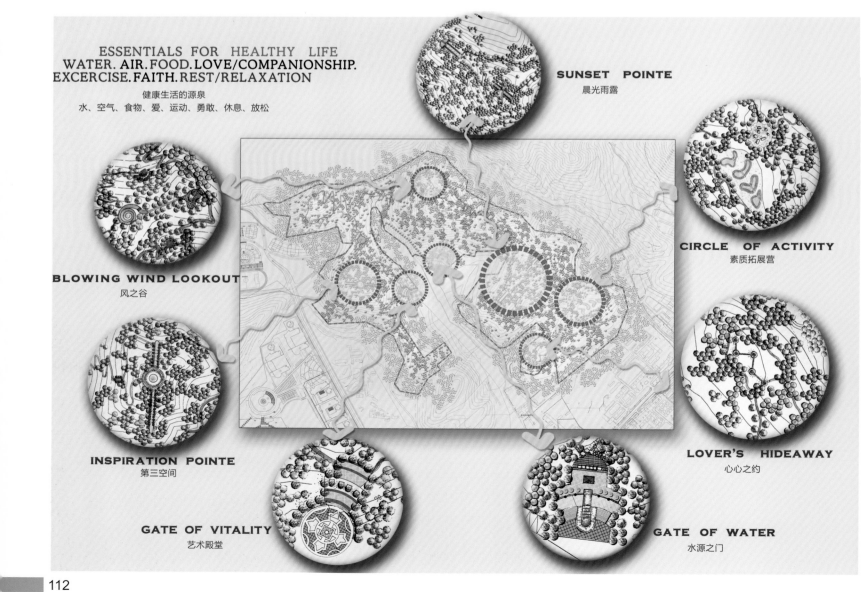

## Overview  项目概况

The project is located in the Donggou Area of Western Dalian Software Park, south to the Software Park Road, connecting Neusoft Group and Neusoft Institute of information in the west, Donggou project in the East and international new town in the southeast, with an area of about 650,000 m². The project is positioned as an international open mountain ecological park with the functions of leisure, culture and fitness to involve more and more people in the mountain experience, providing an good space environment for people to have dialogue with nature; it is built in the type of New York Central Park and will become a representative Green Valley park.

项目位于大连西部软件园区东沟区域，南起软件园路，西接东软公司和东软信息学院，南侧是在建的东沟项目，东南临国际新城，总占地面积约为 65 万 m²。公园功能定位为集休闲、文化、健身为一体的国际化对外开放的山体生态公园，希望更多的人参与到山体中来，为大家提供一个良好的、与自然对话的空间环境；打造具有美国纽约中央公园的效果、具有代表性的绿谷公园。

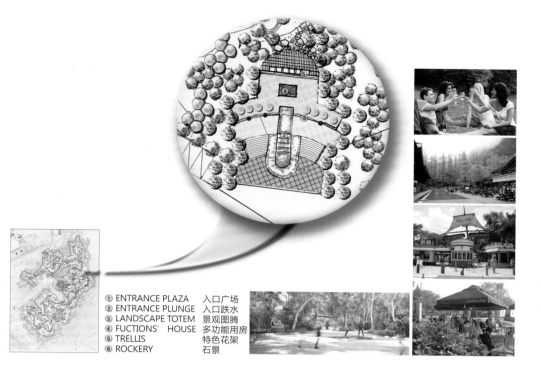

① ENTRANCE PLAZA　　入口广场
② ENTRANCE PLUNGE　　入口跌水
③ LANDSCAPE TOTEM　　景观图腾
④ FUCTIONS' HOUSE　　多功能用房
⑤ TRELLIS　　特色花架
⑥ ROCKERY　　石景

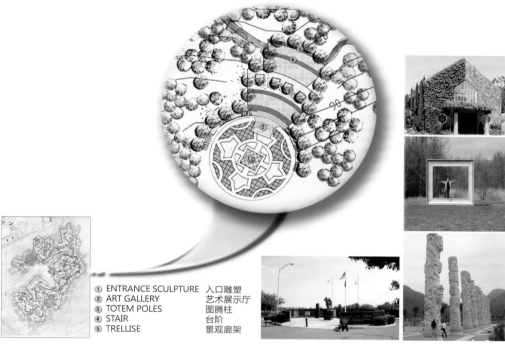

① ENTRANCE SCULPTURE　　入口雕塑
② ART GALLERY　　艺术展示厅
③ TOTEM POLES　　图腾柱
④ STAIR　　台阶
⑤ TRELLISE　　景观廊架

① PLATFORM　　观景台
② SROUND DECK　　景观路
③ ECOLOGICAL BUIDING　　第三空间生态建筑
④ STAIR　　台阶

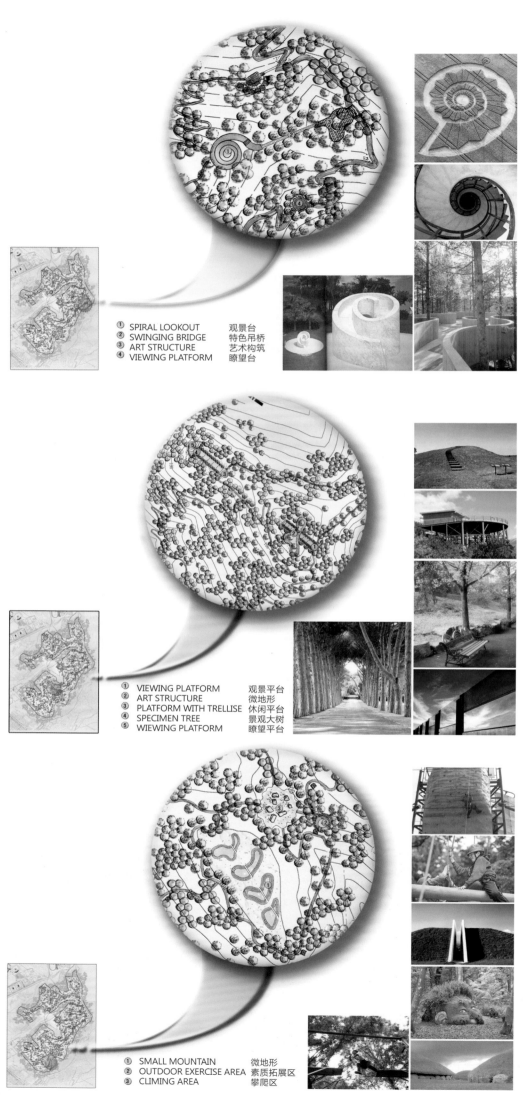

① SPIRAL LOOKOUT 观景台
② SWINGING BRIDGE 特色吊桥
③ ART STRUCTURE 艺术构筑
④ VIEWING PLATFORM 瞭望台

① VIEWING PLATFORM 观景平台
② ART STRUCTURE 微地形
③ PLATFORM WITH TRELLISE 休闲平台
④ SPECIMEN TREE 景观大树
⑤ WIEWING PLATFORM 瞭望平台

① SMALL MOUNTAIN 微地形
② OUTDOOR EXERCISE AREA 素质拓展区
③ CLIMING AREA 攀爬区

## Design Concept 设计理念

Through the optimization of the mountain, it not only improves the mountain and the surrounding landscape environment, but also plays a positive role in the whole environment construction of Dalian. Through the mountain design, it creates a suitable, ecological and natural leisure space and enhances the overall environment of the software park. The mountain park greening should keep minimal damage to the original tree to reflect the natural ecological environment. The design should guarantee the function of extinguishing and protection before the design of the road and its surrounding environment; it keeps the existing fire channel as the main road and optimizes it to be the main channel for extinguishing and protection, then combined with the terrain to make a natural trail along the vertical road. According to the local situation, it can increase the richness of landscape; in the mountain environmental design, the focus should be put on the main entrance landscape, which can provide space for people to linger and also play a guiding role, in addition, according to the mountain situation, it can be considered to design several foot steps trail to reach the mountain top, which can provide protection for people's safety; the design of the mountain road should be optimized based on the protection of the original trees.

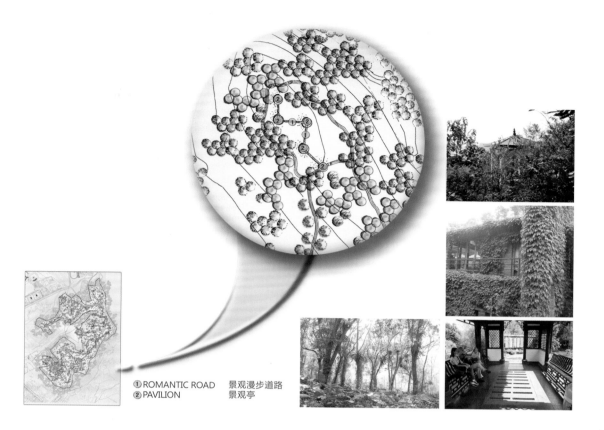

① ROMANTIC ROAD　景观漫步道路
② PAVILION　景观亭

通过对山体的完善优化，不仅对山体本身及周边景观环境有所提升，对大连整个环境建设也起到一定的积极作用。通过山体设计，创造适宜、生态自然的休闲空间，提升软件园区的整体环境。山体公园的绿化应维持对原有树木的最小破坏，以体现自然生态的环境。设计要求在保证消防功能的前提下对道路和道路周边环境进行设计；以现有的防火通道为主路，并加以优化，形成山体主消防通道，结合地形再配以垂直道路修建登山的自然小径。根据现场实际情况，增加景观的丰富性；山体的环境设计中，重点设计主出入口景观，考虑可以让人驻留的空间，同时能起到引导作用，另外依山势情况可考虑设计几条步行台阶小径到达山顶，注意防护，保证对人的安全性；山体的道路铺设要在保护原有树木的前提下进行优化设计。

# TOURISM LANDSCAPE
旅游度假区景观

## MEDITERRANEAN STYLE
地中海风格

## KEY WORDS 关键词

**THEME PARK**
主题公园

**LANDSCAPE NODE**
景观节点

**ECOLOGICAL CHARACTERISTIC**
生态特性

---

Location: Kunming, Yunnan
Developer: Yunnan Overseas Chinese Town Industrial Co., Ltd.
Landscape Design: Shenzhen CSC Landscape Engineering and Design Co., Ltd (CSC)
Partner: CM+CONTEXT Design
Chief Designer: Richard Nugent
Project Director: JOEL
Design team: Zhang Yi, Qin Songze, Sun Ye, Hu Weiqiang, Gao Rui, Zhang Junfeng
Total Planning Area: 7,053,368.6 m²

项目地点：云南省昆明市
开发商：云南华侨城实业有限公司
景观设计：深圳市赛瑞景观工程有限公司
合作单位：CM+CONTEXT 设计
主创设计师：Richard Nugent
项目总监：JOEL
设计团队：张懿 覃松泽 孙烨 胡伟强 高睿 张俊丰
总规划面积：7 053 368.6 m²

# Santorini Hotel Landscape Conceptual Design
圣托里尼酒店景观概念设计

## FEATURES 项目亮点

The beautiful natural surrounding environment and great year-round weather inspire the project to take the Greek islands as the design theme, showing the concept of "humanity and nature".

场地周边美丽的自然环境以及全年宜人的气候激发了项目以希腊式岛屿为主题的设计构思，烘托出"人文和自然"这一概念。

## ▶ Overview 项目概况

Santorini resort village is next to Jiangjia Mountain in the north, Chaoyang Sea in the south and a golf course in the east. Gandhara Ecological Flower Valley extends from Jiangshan Mountain to the resort village and artificial lake in the north. A new residential area will be built around the artificial lake. A fly in the ointment is that the power station is located in the east of the site, so it plants trees to keep the power station out from people's visual field. At the same time, the architectural layout and landscape design take full use of the charming landscape of lakes and mountains of Yangzong Sea.

圣托里尼度假村北靠姜家山，南朝阳宗海，东邻高尔夫球场。犍陀罗生态花谷从姜家山延伸到度假村和人工湖的北面。人工湖周边将建新的住宅区。美中不足的只是电站位于场地的东边，所以景观规划安排种植树林从视觉上来遮盖电站，同时，在建筑布局和景观设计之中充分利用了阳宗海迷人的湖光山色。

View to south 南景观

view to north 北景观

SECTION 1 剖面图 1

SECTION 2 剖面图 2

## ▶ Design Theme 设计主题

The beautiful natural surrounding environment and great year-round weather inspire the project to take the Greek islands as the design theme, showing the concept of "humanity and nature". This design theme leads the entire resort planning, building design and the landscape design of the city and public activity space, foiling the theme of "human and nature".

场地周边美丽的自然环境以及全年宜人的气候激发了项目以希腊式岛屿为主题的设计构思。这个设计主题引导了整个度假村的规划、建筑设计以及城市和公共活动空间的景观设计，烘托了"人文和自然"这一主题。

## Landscape Theme  景观主题

Santorini Hotel is located in the island of an artificial lake; this senior resort hotel bears the unique Aegean Sea type of "island flavor", including the old ramparts, tall castle and beautiful landscape. In order to highlight the theme of Greek Holiday Island, the whole project plans a series of theme park. These theme parks will provide all kinds of activities and island experience for visitors in all ages. These theme parks include: hotel area, bay area, olive garden, "Zuoba" Entertainment Plaza, coastal leisure swimming area, high-level step group, viewing platform and aromatic morning exercise area. In order to emphasize the Greek style, the hotel outside walls and roof garden will plant some vine plants, such as Bougainvillea and Geranium, to make the white building colorful. Along the artificial lake, there are pines, poplars and willows, which set a foil for the hotel.

圣托里尼大酒店坐落于人工湖中的岛屿，这个高级度假酒店具有独特的爱琴海式"小岛风味"，包括古老的城墙、高高的城堡以及优美的景观。为了突出整个希腊式度假岛的主题，整个项目规划了一系列主题公园。这些主题公园将为各个年龄层的游客提供各种各样的活动和小岛体验。这些主题公园包括：酒店片区、海湾片区、橄榄园以及"左巴"娱乐广场、海岸式休闲游泳片区、大台阶组以及观景平台片区、芳香晨练园。为了强调希腊风格，酒店建筑的外立面墙体和屋顶花园将种上叶子花和天竺葵这样的藤本植物，使白色的建筑外观上色彩斑驳；而人工湖沿岸则种植松树、白杨树以及柳树，以烘托酒店。

# TOURISM LANDSCAPE 旅游度假区景观

## CHINESE STYLE 中式风格

**KEY WORDS** 关键词

LANDSCAPE NODE
景观节点

IDEA OF LIFE
养生理念

ECOLOGICAL SPECIALTIES
生态特性

Location: Meizhou, Guangdong
Developer: Guangdong Shanwei Biomedical Group
Landscape Design: Guang Zhou Shi Si Ji Yuan Lin Design Engineering Co., Ltd.
Land Area: 285,347 m²
Floor Area: 28,350 m²

项目地点：广东省梅州市
开 发 商：广东杉维生物医药集团
景观设计：广州市四季园林设计工程有限公司
占地面积：285 347 m²
建筑面积：28 350 m²

# Planning of Nanshufeng King of Medicine Health Preservation Valley

南树峰药王养生谷

## FEATURES 项目亮点

The architects mix the two concepts of Hakka Culture and Medicine Preservation into the landscape pattern and teach visitors through tourism and enable visitors to learn through tourism.

项目将客家文化和药物养生两大理念融入山水格局中，寓教于游、寓学于游。

## ▶ Overview 项目概况

Nanshufeng King of Medicine Health Preservation Valley is located in Songkou Town, Mei County, Meizhou City, covering an area of 285,347 m², with existing roads and living facilities. The present terrain takes the shape of dustpan or palace chair with the valley in middle and ridges on two sides. This terrain belongs to one of the geomantic treasure land according to Chinese traditional culture, which attracts treasure and possesses the advantage of defending itself and a temperament of emperor.

In planning design, architects mix the two concepts of Hakka Culture and Medicine Preservation into the landscape pattern and teach visitors through tourism and enable visitors to learn through tourism. The coming visitors would experience the Hakka Cultural heritage and the South Medicine Culture in the industrial park and understand the traditional farming culture and Chinese profound traditional medicine culture through appreciation and participation.

南树峰药王养生谷位于广东梅州市梅县松口镇，项目占地 285 347 m²，现有基本道路和部分生活设施，呈"簸箕"状，两侧为山脊，中间为山谷，亦可称为"太师椅"状地形，属中国传统文化中认可的招财纳宝、防御自卫、君临天下般的风水宝地。

在规划设计上，设计师把客家文化和药物养生两大理念融入山水格局中，寓教于游、寓学于游，让前来旅游的非客家人感受客家人的文化传承和产业园的南药文化，以观赏和参与的方式了解传统农耕文化和中国博大精深的中医药文化。

Site Plan 总平面图

## Landscape Node 景观节点

Waterfall medicine field appreciation area: the existing terrain displays the high platform with width 100 meters and height 25 meters, which is planned to combine rockwork and waterfall as well as platform-type medicine field. The front side of the slope is designed to be 30-meter wide curtain-type waterfall while the side faces are designed into filiform waterfall in combination of the platform-type medicine field. The entire slope emphasizes on stone-shaping, which stays the focus in the park.

Health and fitness area: utilizing the existing bamboo corridor-Zhangyuan platform to design subtropical outdoor super pool, leisure bar and children's waterpark, etc. for avoiding of summer heat and excising and recreation. Medication rooms are designed at the south side of the bamboo corridor as a supporting facility of the pool in a scattered way to reduce damage to vegetation. The corrie gathers stream and is designed into Wuji Mountainous Landscape Pool for visitors to embrace the original and ecological mountains and forests.

Mountain top hot spring sightseeing area: the existing Tianchi on the high mountain is designed into SPA area, with the mix of pure Tianchi water and classic south medicine to form a multi-type medicine pool, like angelica pool, polygonum multiflorum pool and etc. There are also changing rooms, electric car parking lot for the convenience of sightseeing and bathing. This place is a great resort of mountain overlook, relaxing body and heart and body building.

Comprehensive square ticket office: the square is partially filled and leveled up to 194 meters high, the same height with the existing entrance for easier construction of landscape and forming of an open and ambient entrance.

Hotel residential area: the residential area is basically located on the ridge, focusing on different forms of architectures and multi-type tourist space in the forest, including club hotel, nest type guest room, villa type guest room and health preservation villa, etc.

King of Medicine Valley water landscape sightseeing area: Settling elevation of valley water system is targeted as the main content to improve the value of water and create a five-step fall, namely waterwheel participation area, mountain stream intimacy area, fishing area and water appreciation area. These areas finally join together with Hakka villages water ponds and waterfalls to form a complete mountain and water pattern.

瀑布药田观赏区原有高山平台形状，宽约 100m，高约 25m，拟设计为假山瀑布和台地式药田结合的方式。斜坡面正面设计宽约 30m 的帘状瀑布；侧面结合台地式药田设计丝状飞瀑。整体大斜坡以塑石为主体，成为园区内的视觉焦点。

运动健身区利用原有竹廊"张园"平台，设计亚热带风情露天大泳池、水吧、儿童戏水池等，形成消暑游乐健身的好去处；竹廊南侧设计药疗房作为泳池的配套，散点式以减少对植被的破坏；山凹处聚集溪水并设计为无极山景泳池，体验原生态山林溪水的怀抱。

山顶汤泉游览区利用现状高山天池，设计为药疗 SPA 区，纯净的天池水与正宗的南药融合形成多类型的药物池，如：当归池、何首乌池等；并设计更衣室、电瓶车场以方便游览浸浴，高山远眺、身心放飞、强身健体。

综合广场售票处局部回填平整至 194m，使之与现状入口广场标高平齐，以便于景观建设，并形成开阔大气的入口区。

酒店生活区基本位于山脊，以建筑为主体，有多种形态的建筑及多类型的林间游览空间，包括会所式酒店、鸟巢式客房、别墅式客房、养生别墅等。

药王谷水景观赏区以整理山谷水系标高为主要内容，提高水的利用价值，使水系统形成五级跌落。五级自然跌落划分为水车参与区、溪涧亲水区、钓鱼区、观水区等不同区域，并最终与客家村落水塘及瀑布连为一体，形成完整的山水格局。

## TOURISM LANDSCAPE
旅游度假区景观

## SOUTHEAST ASIAN STYLE
东南亚风格

### KEY WORDS 关键词

**ENVIRONMENTAL PHILOSOPHY**
环保理念

**CULTURAL CONNOTATION**
文化内涵

**LANDSCAPE NODE**
景观节点

Location: Yangjiang, Guangdong
Developer: Yangjiang Nanhu Travel Phoenix Lake International Hot Spring Resort Development Co., Ltd.
Landscape Design: L& A Design Group
Land Area: 4,000,000m²

项目地点：广东省阳江市
开 发 商：阳江市南湖国旅凤凰湖国际温泉度假村开发有限公司
景观设计：奥雅设计集团
占地面积：4 000 000 m²

# The Concept Planning of Yangjiang Spring Resort

阳江温泉度假村

## FEATURES 项目亮点

Themed as ecology and decorated with Southeastern Asian style, combined with local waterscape resource, built a holiday village that possess favorable ecological environment.

项目以生态为主题、以东南亚风情为设计语言，综合当地的水景资源，建设一个具有良好生态环境的度假村。

## Overview 项目概况

Located at southwestern coast of Guangdong Province, Yangjiang is rich in tourism resources with wide variety, high quality and great spatial organization, and near the mountain and by the river. This project covers an area of 4 066 666.67 m² which provides a chance to create large scale and special scenic area. It aims to make an international five-star hot spring resort in Southeastern Asian style by exploring local hot spring resources, which develops local tourist economy, shows local natural ecology and culture and establishes a good ecological environment for the new city.

It is one of Yangjiang's inviting cards to reflect Asian water culture through water-related themes. In addition to interpret local culture and drive the development of local tourism, this design build a high quality service center for welcoming international guests, businessmen and officers.

阳江位于广东省西南沿海，有着丰富的旅游资源，依山傍海，自然旅游资源品种全、品位高、空间组合佳。该项目拥有 4 066 666.67 m² 可用地，为创造规模性景区和特色景区提供了机会。本案设计的目标是打造一个东南亚风情的国际五星级温泉度假村，利用当地的天然温泉资源，打造一系列以水为主题的国际5星级度假休闲天堂，带动当地的旅游经济发展，展示阳江的自然生态与当地文化，为建立良好生态环境的新城提供了机会。

通过与水相关的主题来体现亚洲水文化，成为阳江的一个亮点名片，对此处景观的开发可展示阳江本土文化，带动当地旅游业发展的星级旅游热点，另外，也借此建设了一个接待国际宾客、商人、行政人士等的高级优质服务中心。

### Design Theme 设计主题

Themed as ecology and decorated with Southeastern Asian style, qualitative changes occurred in terms of overall form and function. The unique phoenix-like architectural form corresponds to the theme and style of the resort.

项目以生态为主题、以东南亚风情为设计语言，综合度假村开发。凤凰涅槃，浴火重生，暗喻基地的开发将如山鸡蜕变成凤凰。无论是整体的形态或是使用功能，都是一种质的改变。其建筑形态独特，象征凤凰飞舞，与度假村主题风格相一致。

### Landscape Features 景观特点

The most prominent in the luxuriant vegetation area is that natural water resource is in full use to realize the healthy and environmental concept that the water contains in Asian culture, e.g., drinking tea, organic food, SPA and hot spring. Water is the dominant element in the overall plan with the help of rich water resources, thus waterscapes like lake, stream, wetland and hot spring produced under such a circumstances. Different scenic spots stand a unique space in different scenic areas.

本项目植被茂密，具有景观性，最突出的一点是充分利用天然的水资源来实现亚洲水文化中所蕴含的健康、环保理念。如饮茶、有机食品、SPA、温泉等。利用湖区和小水体所提供的丰富的水资源，把水作为总体规划的主导元素，形成湖、溪流、池塘、湿地、温泉等不同水景。在同一景区中创造不同的景点作为其独特的卖点，体现出公共空间中的个性。

# TOURISM LANDSCAPE
旅游度假区景观

## MEDITERRANEAN STYLE
地中海风格

### KEY WORDS 关键词

- TROPICAL PLANTS 热带植被
- COLOR CONTRAST 色彩对比
- LANDSCAPE NODE 景观节点

Location: Shenzhen, Guangdong
Developer: Shenzhen Hairun Real Estate Development Co., Ltd
Landscape Design: Shenzhen Aode Landscape Design Co., Ltd
Land Area: 5,800 m²

项目地点：广东省深圳市
开 发 商：深圳海润房地产开发有限公司
景观设计：深圳市奥德景观规划设计有限公司
占地面积：5 800 m²

# Shenzhen Aegea · Holiday Mansion
深圳爱琴海·假日公馆

## FEATURES 项目亮点

The designers present the landscape environment and atmosphere through colors, materials, patterns and plants, combined with the local characteristics.

结合项目当地的特色，设计师通过颜色、材料、图案和种植来传达景观的环境氛围。

## Overview 项目概况

The design of the project aims to provide an upscale vacation atmosphere and sense of beauty combining sense of peace, beauty of form and culture to move the visitors.

　　该项目设计的目的是要达到宁静的高端度假氛围，将宁静的感觉、形式美以及文化巧妙地结合起来，使来客被这里美的感觉和氛围所打动。

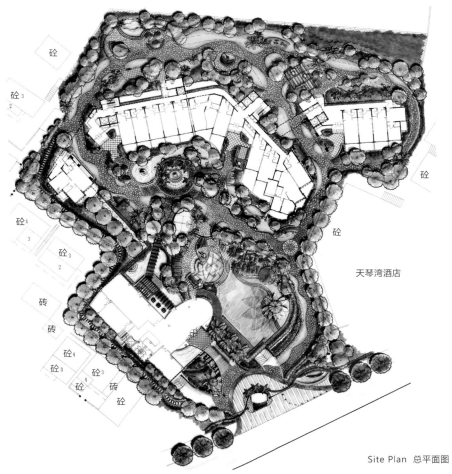

Site Plan 总平面图

### Design Concept  设计理念

To build an impressive, comfortable and unique resort, the designers take every effort to present a hotel environment similar to Aegean style of Greece through the utilization of colors, materials, patterns and plants. It is a popular and charming resort corresponding to the local conditions of Dameisha. The design resembles Santorini landscape style in Greece Island and Aegean, with the elaborate design to integrate with the modern life and the architectures, forming a resort in South China Sea, Shenzhen, China.

为了提供一个难忘的、舒适而又独特的度假胜地，设计师尽量通过颜色、材料、图案和种植来传达一种与希腊爱琴海风格相近的酒店环境，并使之符合大梅沙的当地情况，成为一个颇具吸引力的、受大众欢迎的旅游度假区。设计灵感来自爱琴海和希腊岛如 Santorini 的景观风格，经过设计师的巧妙设计，使其融入现代生活中，并与建筑完美融合，形成位于南中国海海边的度假胜地。

## Landscape Node 景观节点

The swimming pool lies in the center, along with the clubs acting as the landscape focus of the community. Smooth tiles in dark blue color and sandstone floor close to white color decorated in the walls and bottom of the pool have led to a sharp contrast and a strong upscale vacation atmosphere. Tropical characteristic is reflected in the design of tropical plants, and the large palm trees as main plants. The walls are covered by vines, showings a sense of elegance, charming , security and privacy.

游泳池位于中心区域，与会所形成全区的景观焦点。泳池墙壁和底部用深蓝色光滑的瓷片与接近白色的砂岩地面形成强烈对比，打造出浓郁的高端度假氛围。热带植被的设计具有热带特色，大型棕榈树成为主景植物。墙体被攀爬植物隐藏起来，给人以优雅迷人及安全隐蔽的感觉。

## TOURISM LANDSCAPE
旅游度假区景观

## MODERNISM STYLE
现代主义风格

### KEY WORDS 关键词

ECOLOGICAL CHARACTERISTIC
生态特质

THEMATIC FEATURE
主题特色

LANDSCAPE NODE
景观节点

Location: Hefei, Anhui
Landscape Design: DDON Associates
Landscape Area: 3,748,000 m²

项目地点：安徽省合肥市
景观设计：笛东联合（北京）规划设计顾问有限公司
景观面积：3 748 000 m²

# Hefei Lakefront Forest Park
合肥滨湖森林公园

## FEATURES 项目亮点

Maintaining the integrity of original vegetation, using scientific methods to promote ecological cycle and create a leisure and tourism resort destination featuring ecology first, wetland recreation, forest experience and exhibition education.

保持原生态植被完整性，通过科学手段促进生态循环，以打造生态优先，集湿地游憩、森林体验、展示教育于一体的"休闲旅游度假目的地"。

> **Overview** 项目概况

The project is located to the south of Jiazi River, north of Huanhu North Road, east of Choahu South Road and west of Nanfei River. North-south Shiwuli River runs through the site. To a larger extent, it is adjacent to urban expressway Chongqing Road and east-west expressway Fangxing Road, and the main landscape road, Huanhu North Road, runs across the site. It is just 15 km away from Luogang Airport, boasting a very convenient traffic with dense road network around.

项目北依甲子河，南达环湖北路，西连巢湖南路，东临南淝河，十五里河南北贯穿。基地西侧紧依城市快速路重庆路，东西向城市快速道路方兴大道、环巢湖主要景观道路环湖北路横穿基地，周边道路网密集，距离骆岗机场仅 15 km，交通极为便利。

### Design Goal 设计目标

It is committed to create a leisure and tourism resort destination featuring ecology first, wetland recreation, forest experience and exhibition education. It keeps integrity of original vegetation and adopts scientific methods to promote ecological cycle. It strictly controls and improves the construction, creates prominent features and plots out the touring route in accordance with the characteristics of existing resources.

项目以打造生态优先，集湿地游憩、森林体验、展示教育于一体的"休闲旅游度假目的地"为设计目标。项目保持原生态植被完整性，不破坏，不拆除，并通过科学手段促进生态循环，严格控制建设量，在场地现有资源和条件的基础上进行改造提升；突出特色，营造亮点，根据资源特点建立丰富的旅游产品体系及游览路线。

### Landscape Planning 景观规划

The project plan is centered on four things: node, corridor, interface, development zone. The whole site is divided into three parks and twelve areas. Three parks are interactive tour park, eco-farm experience park and forest interactive recreation park. The areas are corporate culture base, wetland interactive experience area, wetland theme activity area, commercial plaza and entrance service area, wetland amusement recreation area, waterfront business district, waterfront dining area, wetland ecological education area, forest recreation experience area, forest conservation area, tourist area of farmland and ecological resort.

The wetland irrigation system is improved to become a combination of irrigation drainage system and circulatory system. There are mainly wetland water cycle and forest water cycle in this site. Wetland water cycle involves in Shiwuli River, Chaohu, Jiazi River and Nanfei River, and a regulator is set at the mouth of Shiwuli River to maintain the internal water cycle, the tributary Jiazi River flows to Nanfei River and then into Chaohu River. Forest water cycle forms a peripheral ring canal internally based on the original canal, and forms a shaped circulatory system with the former east-west main canal. Water transfer in different levels guarantees the sufficient water cycle for the whole park.

项目从"节点、廊道、界面、开发区"四个方面规划。整个项目将地块分为"三园十二区"。"三园"包括湿地互动游览公园、生态农庄观光体验园、森林互动游憩公园。"十二区"包括企业文化基地区、湿地互动体验区、湿地主题活动区、商业广场及入口服务区、湿地游憩区、滨水商业区、滨水特色餐饮区、湿地生态教育区、森林游憩体验区、森林生态保护区、农田观光区、生态度假区。

湿地灌溉系统整体形成排涝灌溉系统和循环系统一体化设计。基地主要有湿地水循环和森林水循环。湿地水循环主要是在十五里河入巢湖处设节制闸，内部单独进行水循环，支流甲子河流向南淝河最终汇入巢湖。森林水循环内部在原有水渠基础上整理形成外围环型水渠，与原有东西向主渠形成日字形主循环系统，沿干渠分段设计泵站形成分级调水，通过高程整理结合泵站坝体保证整个园区形成充分的水循环，最后由东林站排水入南淝河；森林公园外部形成湿地水循环。

# TOURISM LANDSCAPE
## 旅游度假区景观

## CHINESE STYLE
## 中式风格

### KEY WORDS 关键词

- LANDSCAPE NODE 景观节点
- CULTURE FEATURE 文化特色
- ECOLOGICAL CHARACTERISTIC 生态特质

Location: Taizhou, Jiangsu
Developer: Overseas Chinese Town
Landscape Design: Shenzhen DongDa Landscape Design Co.,Ltd.
Chief Designer: Zhou Yongzhong, Gu Mulan, Yang Yi, Yu Liang
Landscape Area: 49,000m²

项目地点：江苏省泰州市
开 发 商：华侨城集团
景观设计：深圳市东大景观设计有限公司
主设计师：周永忠 顾慕兰 杨沂 俞亮
景观面积：49 000 m²

# Taizhou OCT Hot Spring SPA Landscape

## 泰州华侨城温泉SPA景观

### FEATURES 项目亮点

This design blends local ecological and cultural elements and the actual environment of the site, so that people can enjoy the peace and quiet melted in the nature while enjoy the hot spring SPA, experiencing local peculiar culture and traditional complex.

设计细节融合当地生态、文化元素并结合场地的实际环境，使人们在享受温泉SPA的同时，亦能享受被自然生态所包容的那份清净，体会当地特殊的文化韵味及传统情结。

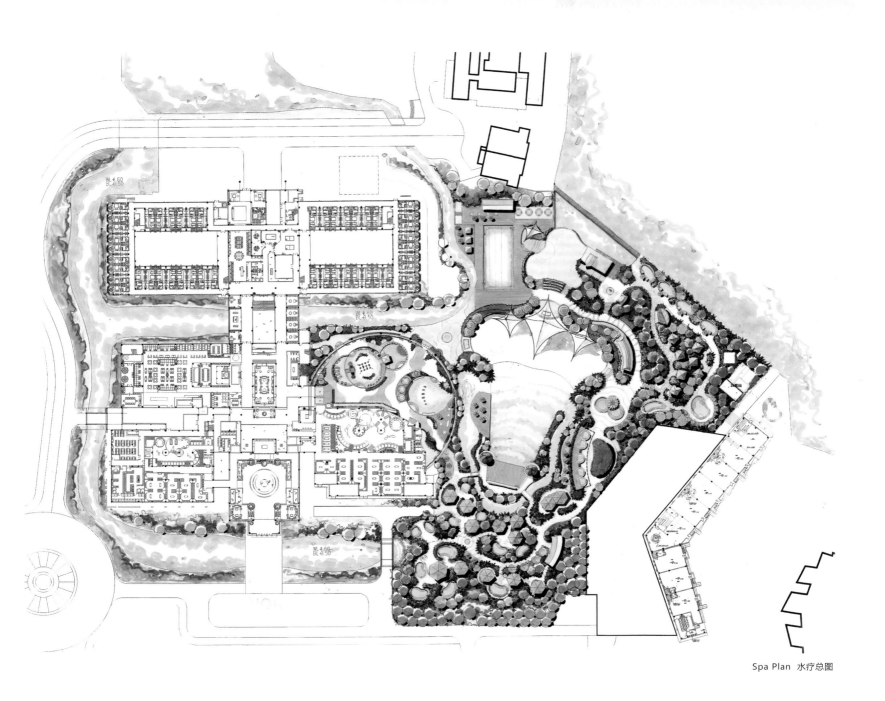

Spa Plan 水疗总图

### Overview 项目概况

This project is located in the west of Jiangyan Qinhu Scenic Area which is a typical ecological wetland and has natural hot spring resources. It is a key subproject for Taizhou OCT Ecotourism integrated project and a tipping point for OCT's business in Taizhou.

　　泰州温泉 SPA 项目位于江苏省姜堰市溱湖风景区西片区，是典型的湖荡生态湿地，并拥有天然的温泉资源，为泰州华侨城湿地生态旅游综合项目一期重点打造的子项目，也是泰州华侨城项目开发的引爆点。

### Design Concept 设计理念

Design philosophy is mainly reflected in the ingenious integration of local characteristic culture and landscape details. This design blends local ecological and cultural elements and the actual environment of the site, so that people can enjoy the peace and quiet melted in the nature while enjoy the hot spring spa, experiencing local peculiar culture and traditional complex.

　　设计理念集中体现在将当地的特色文化遗产巧妙地融合到景观细节当中。整个项目中，设计方案解决了设计细节结合当地生态及文化元素这个课题，利用当地文化的特色结合场地的实际环境使人们在享受温泉 SPA 的同时，亦能享受被自然生态所包容的那份清净，体会当地特殊的文化韵味及传统情结。

## Cultural Identity of Landscape 景观文化特性

Lixia River, where the project located, is a natural wetland zone. Adhering to the wetland pattern, designers try to follow and restore the characteristics of original wetland landscape plant to create a mysterious and green landscape experience. Compared to other Spa areas, this one is much more natural and ecological. Lotus is the main wetland plant, whose image is also embedded in various landscape elements such as pattern, sculpture, lamp and floor decoration, etc. to highlight the regional ecological and cultural characteristics. In addition, reeds and birds area also imported to increase overall wetland ecosystem atmosphere.

Traditional Chinese pavilion blends the essence of the culture of Lixia River and Yangtze River Delta, which serves as shelter and creates a cultural conception of ancient times. Taizhou is the land of traditional opera which plays a key role in connecting all the design details.

Some hot spring pool are named after the ancient poetry to portray beautiful sceneries the scholars painted, such as Yi Jiang Nan (Memories of the South), Cai Sang Zi (Picking Mulberries), Die Lian Hua (Love of Butterfly) and Huan Xi Sha (Sand of Silk-washing Stream), etc. In addition, some are designed in accordance with the people's love for the number of 5.

本项目所在位置属于里下河地区，是一个天然的湿地地带。设计秉承湿地格局，尽量沿用及还原湿地景观植物特色，营造曲径通幽、绿意满盈的景观体验，较其他温泉SPA区更为自然及生态。湿地植物以荷花为主，利用荷花的造型延展于各个景观设计元素中，包括各种不同图案的组合构图、小品雕塑、灯具及地面铺装等，以突出地域生态和文化特色。另外，利用依附于湿地属性的物质元素，如芦苇、鸟类等，增加整体的湿地生态氛围。

中式古亭融合了里下河文化及江南水乡文化的精髓，传统特色较为明显。设计中利用古亭结合温泉泡池，既可纳凉避风，又营造了思古徜徉的文化意境。泰州乃戏曲之乡，在项目中，戏曲作为情感主线之一，串联融入设计细节，在浮雕小品乃至规划布局中皆可看到戏曲文化的影子。

设计努力描绘文人骚客描画的江南美景，每一处温泉泡池均根据诗词来命名，如忆江南、采桑子、蝶恋花、浣溪沙等。并利用当地人对于5这个数字的钟爱，设计了诸如五子浴（生姜、女贞、白蒺藜、白芷、朱砂）、五音浴（宫、商、角、徵、羽）、五果浴（苹果、木瓜、柠檬、菠萝、椰子）、五花浴（茉莉、菊花、桂花、玫瑰、郁金香）等泡池。

# Urban Planning
# 城市规划

Regional Features
地域特征

Functional Landscape
景观功能

Culture Experience
文化体验

## URBAN PLANNING
城市规划

## MODERN STYLE
现代风格

### KEY WORDS 关键词

**PRIMITIVE ECOLOGY**
原始生态

**STRIP LANDSCAPE**
带状景观

**LANDSCAPE NODE**
景观节点

Location: Jiangmen, Guangdong
Developer: The Town Government of Hecheng, Jiangmen City
Landscape Design: Shenzhen Chengbang Landscape
Total Land Area: 567,950.5m²
Total Floor Area: 1,298.5m²
Greening Ratio: 99.1%

项目地点：广东省江门市
开 发 商：江门鹤城镇政府
景观设计：深圳城邦园林景观工程有限公司
总用地面积：567 950.5 m²
总建筑面积：1 298.5 m²
绿 化 率：99.1%

# Kunlun Rainbow Landscape Design in Hecheng Town, Jiangmen City

江门市鹤城镇昆仑彩虹景观设计

## FEATURES 项目亮点

This landscape design take advantage of the strength, creating a greenway landscape attraction of distinctive layout and rich vegetation along Kunlun Mountain's winding lines.

设计借势造景，沿着昆仑山曲折的线路打造了一个层次鲜明、植被丰富的旅游绿道景观。

## ▶ Overview 项目概况

The project is in Kunlun Mountain, Hecheng Town, Jiangmen City, Guangdong Province. The atmosphere here is simple, steep mountains and wide field of horizons. The focus is not just the greenway of Grand Kunlun Mountain, but also looking out the whole Pearl River Delta to build a grand Kunlun Celestial Mountain. Being positioned as a greenway of mysterious celestial mountain, the greenway connects every landscape group.

项目位于广东江门市鹤城镇昆仑山，环境氛围朴实，原始风味浓重，山势陡峭，视野宽广。其着眼点不仅仅是大昆仑山绿道、而是放眼珠三角，打造大昆仑仙山。定位为打造神秘的仙山绿道，通过绿道串联起一个个景观组团。

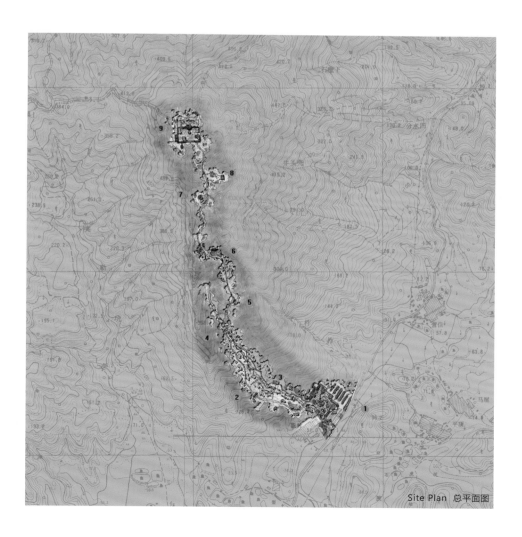

Site Plan 总平面图

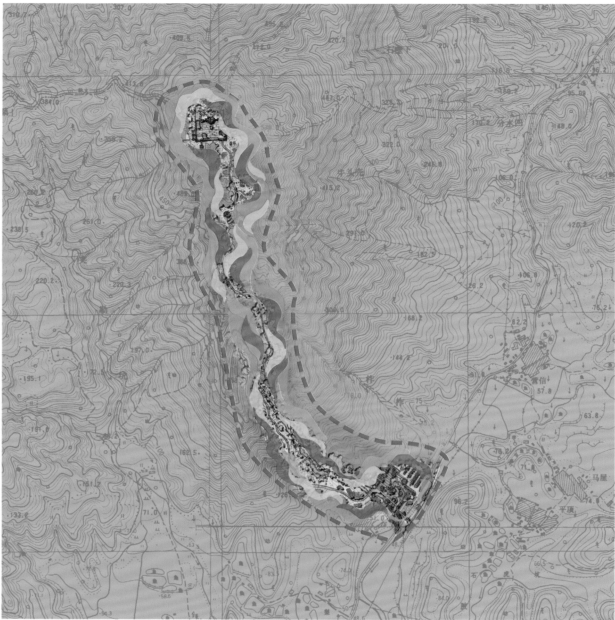

春
夏
秋
冬

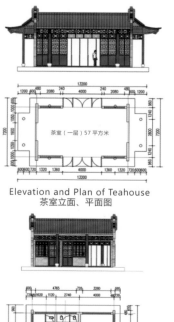

Elevation and Plan of Teahouse
茶室立面、平面图

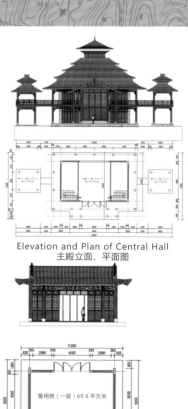

Elevation and Plan of Central Hall
主殿立面、平面图

Elevation and Plan of Washroom and Store
洗手间及小卖部立面、平面图

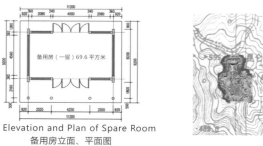

Elevation and Plan of Spare Room
备用房立面、平面图

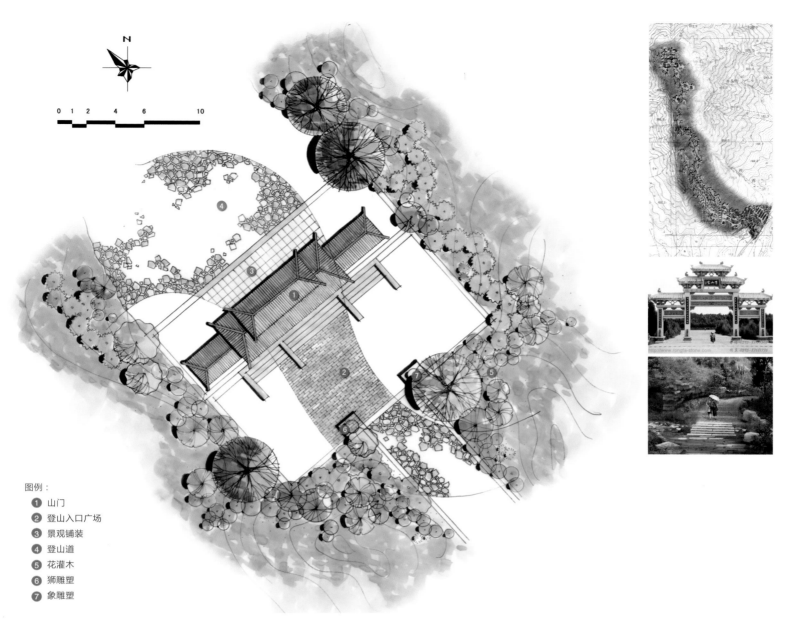

图例：
1. 山门
2. 登山入口广场
3. 景观铺装
4. 登山道
5. 花灌木
6. 狮雕塑
7. 象雕塑

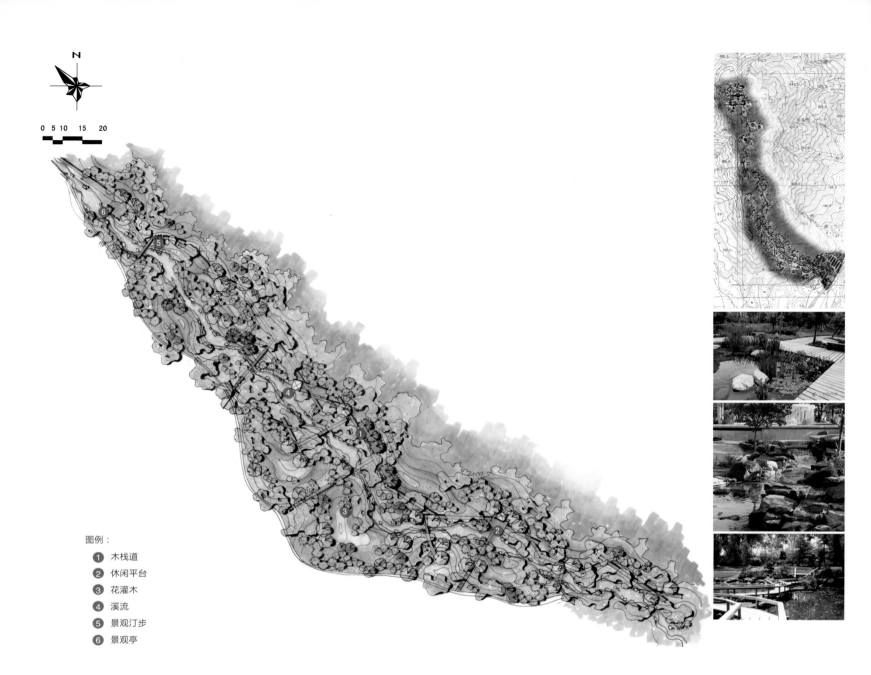

图例：
1. 木栈道
2. 休闲平台
3. 花灌木
4. 溪流
5. 景观汀步
6. 景观亭

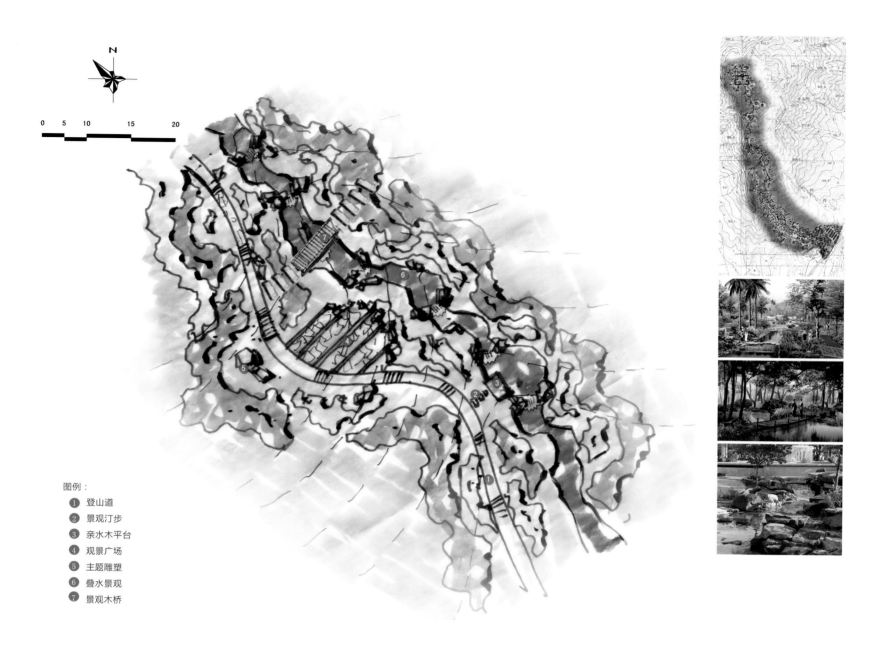

图例：
1. 登山道
2. 景观汀步
3. 亲水木平台
4. 观景广场
5. 主题雕塑
6. 叠水景观
7. 景观木桥

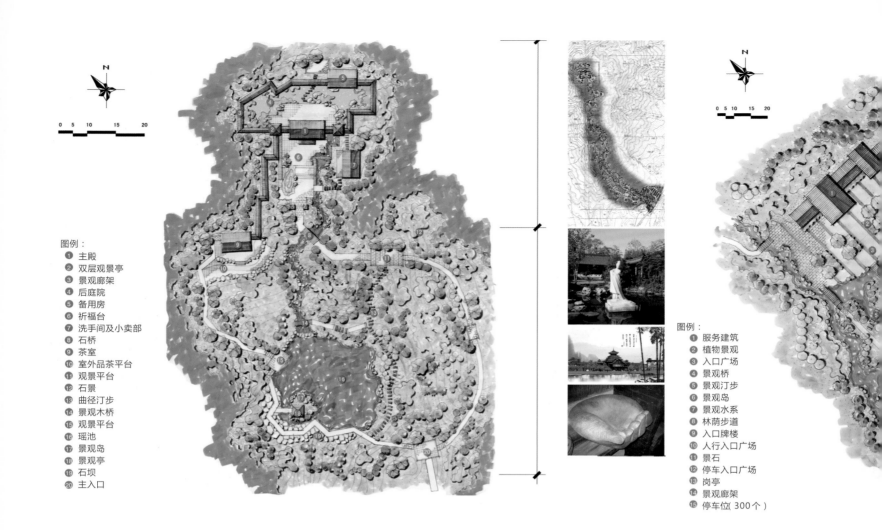

图例：
1. 主殿
2. 双层观景亭
3. 景观廊架
4. 后庭院
5. 备用房
6. 祈福台
7. 洗手间及小卖部
8. 石桥
9. 茶室
10. 室外品茶平台
11. 观景平台
12. 石景
13. 曲径汀步
14. 景观木桥
15. 观景平台
16. 瑶池
17. 景观岛
18. 景观亭
19. 石坝
20. 主入口

图例：
1. 服务建筑
2. 植物景观
3. 入口广场
4. 景观桥
5. 景观汀步
6. 景观岛
7. 景观水系
8. 林荫步道
9. 入口牌楼
10. 人行入口广场
11. 景石
12. 停车入口广场
13. 岗亭
14. 景观廊架
15. 停车位（300个）

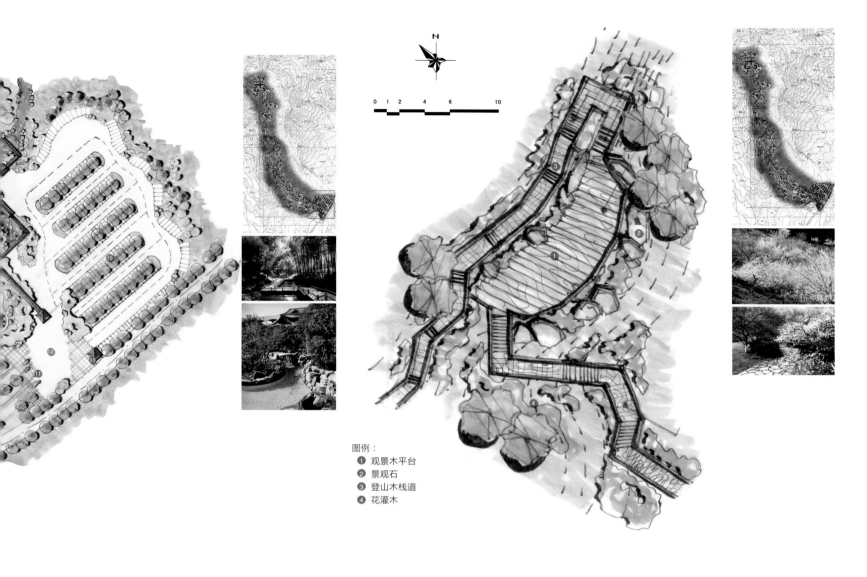

图例：
1. 观景木平台
2. 景观石
3. 登山木栈道
4. 花灌木

### ▶ Planting Design  绿植设计

The spring plants is mainly red eryt hrina crista-galli Linn and red bombax ceiba, among which alternates pink magnolia liliiflora as well as embellishes with Golden Trumpet-tree. Then the additional peach trees and plum trees is helpful to create a plant landscape with various color and layout. When it comes to the summer plants, delonix regia, Lagerstroemia speciosa, and spathodea campanulata are the framework, complemented with evergreen trees to make a landscape of greenery safflower. Next, the autumn plant design is based on tree with red, yellow leaves, as well as interspersed with National Day flowers, ceiba insignis, trident maple and Aoihana to enrich the color layout, so as to reach the effect that flowers, leaves, colors all have different shape which provides fresh feeling for tourist. Last is the winter plant design, the shrubs bloom in winter, such as azalea, rhododendron pulchrum, rhododendron rufum, wintersweet etc, is used to create a visual shock that the colorful flowers bloom in cold winter making tourist feel heart-warming.

春季植物的设计以红色系的鸡冠刺桐、木棉为主，穿插以粉色的玉堂春，同时点缀黄花风铃木，增设主题植物桃、李等营造色彩层次丰富的植物景观。夏季植物以凤凰木、大叶紫薇、火焰木等为骨架，穿插以常绿乔木，打造花红叶绿的景观。秋季植物设计，乔木以红色、黄色以及色叶植物为主，配以国庆花、美丽异木棉、三角枫、蓝花等增加色彩的层次，达到花、叶、色形态各异的效果，给人以神清气爽之感。冬景的植物设计，以映山红、锦绣杜鹃、毛杜鹃、腊梅等冬季开发灌木作搭配，营造冬寒花开暖人心的视觉冲击。

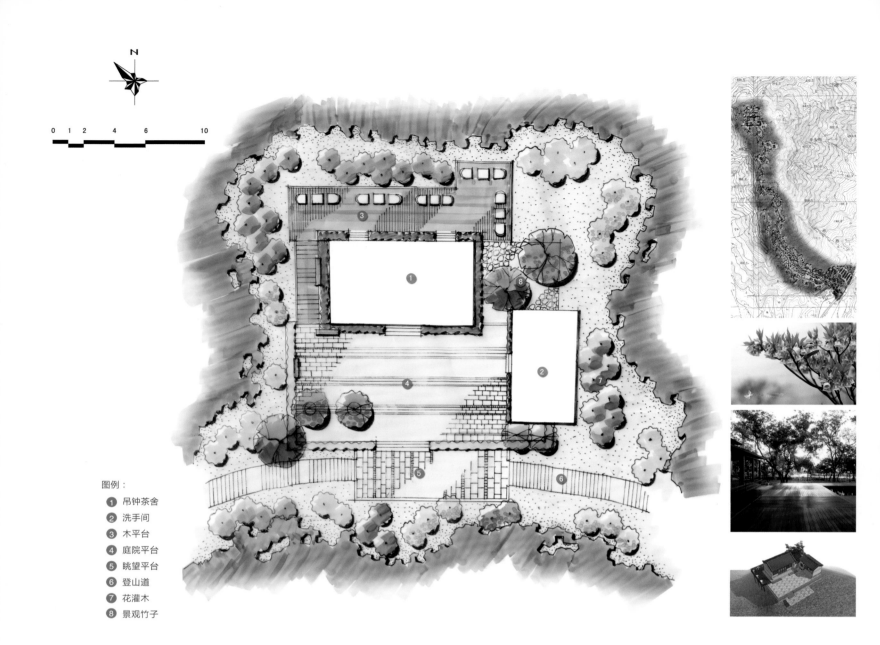

图例：
1. 吊钟茶舍
2. 洗手间
3. 木平台
4. 庭院平台
5. 眺望平台
6. 登山道
7. 花灌木
8. 景观竹子

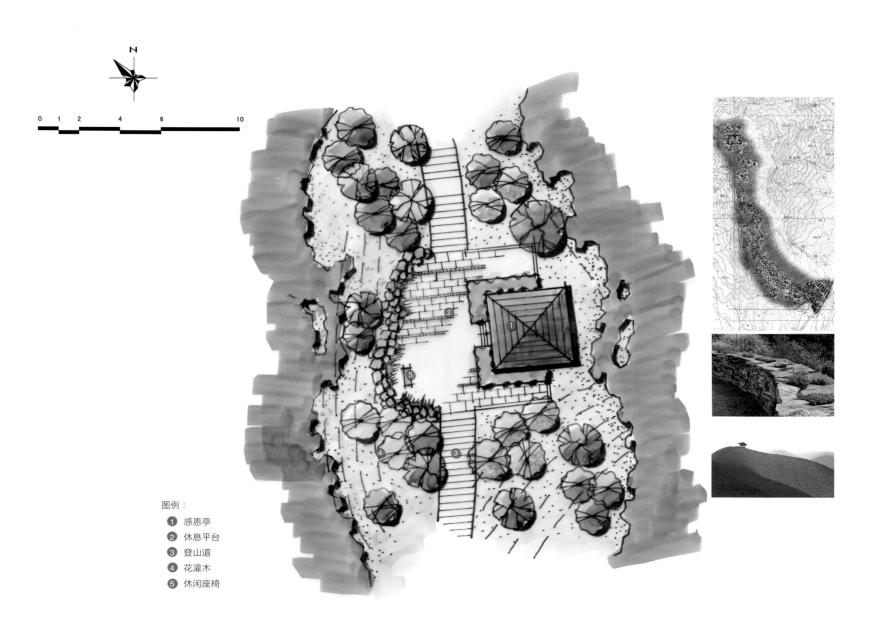

图例：
1. 感恩亭
2. 休息平台
3. 登山道
4. 花灌木
5. 休闲座椅

# URBAN PLANNING
# 城市规划

## CHINESE STYLE
## 中式风格

**KEY WORDS** 关键词

CULTURAL THEME
文化主题

ECOLOGICAL CHARACTERISTIC
生态特质

LANDSCAPE NODE
景观节点

Location: Luzhou, Sichuan
Developer: Luzhou Lijia Properties Co., Ltd.
Landscape Design: Shenzhen DongDa Landscape Design Co., Ltd.
Total Land Area: 2,186,700m²
Construction Area: 798,000m²
Landscape Area: 1,398,700m²

项目地点：四川省泸州市
开 发 商：泸州市利嘉置业有限公司
景观设计：深圳市东大景观设计有限公司
总用地面积：2 186 700 m²
建设用地面积：798 000 m²
景观面积：1 398 700 m²

# Luzhou City Xueshishan Park
# 泸州市学士山公园

## FEATURES 项目亮点

With the idea of creating a "green heart" of the city, Xueshishan Park will be a comprehensive park integrating the functions of commerce, recreation and science education.

以创造城市"绿心"、营造最具"活力"的城市公园为理念，融合商业、休闲、科教等功能，共同打造泸州市大型城市综合公园。

## Overview 项目概况

The site locates in the northeast of downtown Luzhou with convenient traffic conditions. As the well-preserved green land within the city, it features the unique resources such as mountain, forest, waters, historic neighborhood, TV Tower, etc.

项目位于四川省泸州市中心城区东北面，交通便利，是城区保留得最完好的绿地。山林水系资源丰富，还有历史街区、电视塔等特色资源。

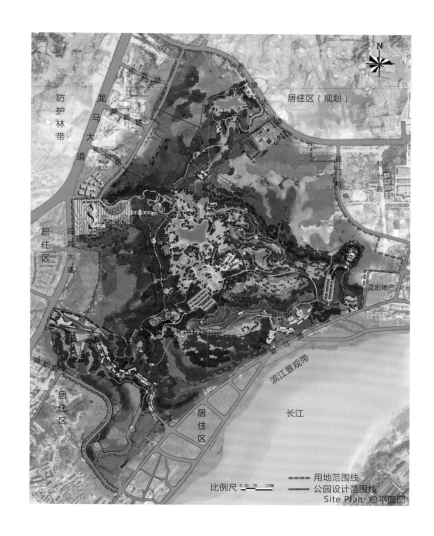

Site Plan 总平面图

### Design Concept 设计理念

With the idea of creating a "dynamic" urban park, it is envisioned to the "green heart" of the city. Xueshishan Park will be a comprehensive park integrating the functions of commerce, recreation and science education to show the history and culture of Luzhou City.

项目以创造城市"绿心"、营造最具"活力"的城市公园为理念，融合商业、休闲、科教等功能，展现泸州当地的历史文化，共同打造泸州市大型城市综合公园。

## Landscape Construction 景观营造

Landscape design of this park pays attention to embody local culture and art with dynamic shapes, different colors and varied forms. Wine Culture, Red Culture, Folk Culture and Green Culture of the park are well explained, for example, there is Wind Cultural Square, wine-bowl building, etc. In terms of eco design, it respects the original topography, and tries to protect and make use of the existing ecological resources. Environment-friendly materials are applied to create eco and green buildings. The landscape design not only protects the original green resources but also diversifies the green spaces with different colors and forms, providing great spaces for different activities.

公园景观设计着重体现当地文化艺术的内涵，运用造型动感、色彩形式丰富的设计手法，综合体现公园酒文化、红色文化、民俗文化、绿色文化等主题，如酒文化广场、大酒碗建筑等。生态环保方面整体设计在原地形的基础上，保护和利用原有生态资源，利用环保材料将功能服务建筑建设成生态建筑、环保建筑。公园景观设计在保护原有绿化的基础上，局部强化绿化的色彩及造型，营造丰富的空间活动场地。

# URBAN PLANNING
城市规划

## MODERN STYLE
现代风格

## KEY WORDS 关键词

### ECO NATURE
生态特质

### LANDSCAPE CORRIDOR
景观廊道

### LANDSCAPE NODE
景观节点

Location: Chizhou, Anhui
Developer: Anhui Zhongkatong Animation Co., Ltd.
Landscape Design: Shenzhen DongDa Landscape Design Co., Ltd.
Total Planning Area: 2,000,000m²

项目地点：安徽省池州市
开 发 商：安徽中卡通动漫有限公司
景观设计：深圳市东大景观设计有限公司
总规划面积：2 000 000 m²

# Jiuhuashan Animation Theme Park
九华山动漫主题乐园

## FEATURES 项目亮点

With the focus on "one park, two lines, three points and two axes", it combines amusement park with landscape park, providing eco experience and amusement experience at the same time.

设计重点打造"一园、两线、三点、两轴"的景观模式，将乐园与公园结合，打造生态体验与游乐休闲相结合的主题乐园。

### Overview 项目概况

The park is designed with three functional areas: the amusement area with the theme of "sky, earth and human beings", the eco landscape experience area and the water world.

九华山动漫主题乐园的项目设计分为天、地、人主题游乐区，生态景观体验区，神游水世界三大功能区。

## Design Goal 设计目标

The design focuses on creating "one park, two lines, three points and two axes". One park refers to the development center, two lines form the impression interface; three points are the landscape nodes; two axes mean the landscape visual corridors.

设计重点打造"一园、两线、三点、两轴"的景观模式。一园——发展触媒；两线——形象展示界面；三点——标识节点；两轴——视线景观廊道。

185

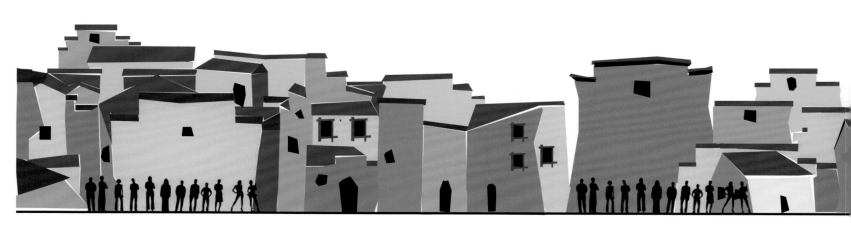

## Landscape Theme 景观主题

With respect to the original topography, it combines amusement park with landscape park, trying to provide eco experience and amusement experience at the same time. The forest is transformed into an eco culture experience zone, preserving the original landform. It encourages rainwater collection and reduces rainwater storage in the grass bunkers or lawns. In this development, landscape, ecology, design and construction are combined and considered together to create a unique animation theme park.

设计尊重地形地貌，将乐园与公园结合，打造生态体验与游乐休闲相结合的主题游，对林相进行改造，构建以山林地为特征的生态文化体验区，强化地理格局记忆的保留。鼓励雨水收集以亲近自然，使在硬地降落面的雨水流入草坑或草坪洼地临时存储。在整个项目中将景观、生态、设计、施工等综合考虑，创造与众不同的动漫主题乐园。

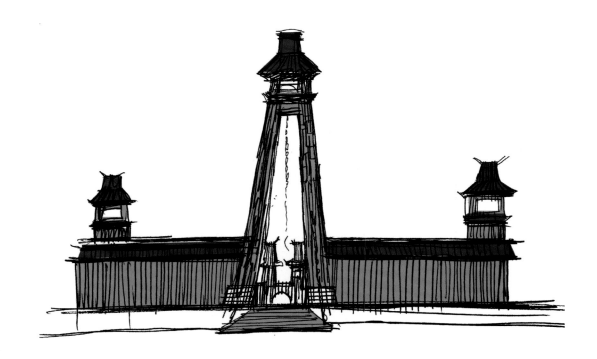

# URBAN PLANNING
## 城市规划

**CHINESE STYLE** 中式风格

# China (Chongqing) International Garden Expo
## 中国(重庆)国际园林博览园

**KEY WORDS** 关键词

- ECO QUALITY 生态特质
- REGIONAL FEATURE 地域特色
- LANDSCAPE NODE 景观节点

## FEATURES 项目亮点

With the idea of "telling the story of Chongqing City", it takes advantage of the natural topography and creates a beautiful landscape environment.

总体规划本着"述说巴渝文化"的理念，利用原址自然地貌，营造出一个依山傍水、自然优美的总体环境。

Location: North New District, Chongqing
Developer: Chongqing Government, Chongqing Municipal Bureau of Gardens
Landscape Design: Shenzhen DongDa Landscape Design Co.,Ltd.
Total Floor Area: approximately 2,000,010m²

项目地点：重庆北部新区
开 发 商：重庆市政府、重庆市园林局
景观设计：深圳市东大景观设计有限公司
总建筑面积：约 2 000 010 m²

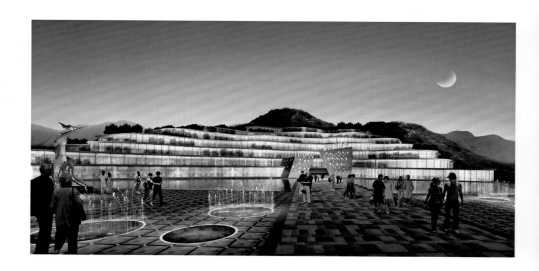

## Overview 项目概况

The EXPO chose its site in Longjing Lake Area of Chongqing North New District with a total area of 2,000,010 m². It includes Longjing Lake and many hills around the lake. The altitude ranges from 276 m to 421 m.

重庆园博园的选址点位于重庆主城核心区——北部新区龙景湖片区，园区占地面积约 2 000 010 m²，由龙景湖和周围的多个小山丘组成，海拔高度在 276 m 至 421 m 之间。

## Design Concept 设计理念

With the idea of "tell the local culture of Chongqing", it takes advantage of the topography and creates a place with hills, waters and beautiful landscapes. The "Fire Phoenix" is the totem of the park to symbolize the "transformation" of the garden expo. After the expo, it will be a park with the theme of "ecology, recreation and tourism" to enrich people's leisure time.

总体规划本着"述说巴渝文化"的理念，利用原址自然地貌，营造出一个依山傍水、自然优美的总体环境。用"火焰凤凰"作为园区图腾，塑造"转型化"的园博会。会展时期的欢乐园博会，会后将形成一个以"生态、休闲、旅游"为主题的公园来补充重庆城区内的市民对健康休闲游的需求。

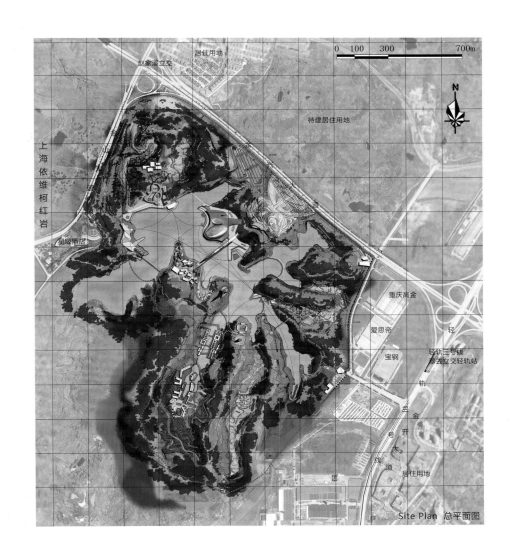

Site Plan 总平面图

## Landscape Planning 景观规划

According to the overall planning, there are gardens of South China, North China, Southeast China and the Three Gorges area, the international gardens and the masters' gardens. The design embodies the local characteristics and reflects Chongqing's culture. The boutique garden with the "dragon back rock" shows the wharf culture and indicates that Chongqing is "the city of thousand bridges". In addition to the above designs, to achieve a systematic construction, it also provides many suggestions on ecology, science and technology, plants, materials, labels, facilities, transformation and sustainable development.

总体规划设立中国江南、北方、岭南、三峡园林展示区以及国际园林展示区、大师园区。反映地方特色、展现巴渝文化，结合地形高差和保护现状"龙脊"石体现码头文化的巴渝精品园，结合高差反映重庆"千桥之城"。除以上规划和设计外，为保证园区建设更加系统，还针对生态、科技、植物、材料运用、标识、设施小品以及会后转换与可持续发展等提出了规划建议。

## Landscape Features 景观特色

Landscape at the main entrance is characteristic with the unique elements of Chongqing City – "hills and terraced fields". At the 33,000m² entrance area, there is a meeting square and a tourist center. The design of the tourist center is inspired by different altitudes and terraced fields of this mountain city, and creates an eco building complex. Buildings are built following the topography to combine with the mountains perfectly. Glass, bamboo and wooden walls, roof plantation are applied to reduce energy consumption. Terraced fields and rocks combine with the waterscapes in the entrance square, showing the unique landscape of Chongqing City.

园博园的主入口景观极具特色。"山地、梯田"为山城重庆的独有的记忆和特征，设计师以此概念作为出发点来表现重庆园博园的主入口形象。主入口区由面积为33 000 m²的集散广场及综合游客中心组成。综合游客服务中心建筑形态来源于山城丰富变化的高差及根据高差而生的梯田，整体概念为"梯田"和"岩石"的生态综合建筑。建筑依靠现有的裸露山体而建，完美地融合地形山体。建筑使用低能耗的玻璃、竹木墙，采用屋顶种植技术，以减少建筑能耗及总体热辐射能量。"梯田"和"岩石"与入口广场的大面积水景呼应，体现重庆城市中"山水相映"的特色地貌。

# Public landscape
# 公共景观

Plant Landscape
植物造景

Landscape Theme
景观主题

Public Space
公共空间

# PUBLIC LANDSCAPE
## 公共景观

## CHINESE STYLE
## 中式风格

**KEY WORDS** 关键词

**LOCAL CHARACTERISTIC**
地方特色

**RUSTIC STYLE**
乡村风情

**LANDSCAPE NODE**
景观节点

# South Bank Riverside Scenic Zone of Zhengshui River, Hengyang
## 衡阳蒸水河南岸滨江风光带

### FEATURES 项目亮点

Urban modernization, native landscape and folk culture are integrated in the project to generate the poetic dwelling environment.

项目把城市现代化建设、乡土景观和民俗文化结合在一起，创造出"诗意栖居"的意境。

Location: Hengyang, Hunan
Developer: Hengyang Yinrun Construction Investment Co., Ltd.
Landscape Design: Shenzhen Sairui Landscape Engineering Design Co., Ltd.
Chief Designers: Ye Qiang, Ding Jiong
Project Director: John Chaw
Project Team: Daisy Peng, Betty Lee, Jia Li, Kevin Zhao, Lucy Cao, Chunhong Ming, Willa Wu, Peng Liu, Pene Pang, Bo Chen, Meifang Sun, Huifang Yi, Cindy Yang
Area: 38,0000 m²

项目地点：湖南省衡阳市
开 发 商：衡阳市银润建设投资有限公司
景观设计：深圳市赛瑞景观工程设计有限公司
主创设计师：叶强 丁炯
项目总监：邹炯
项目团队：彭妍 李婧 庞美赋 李佳 赵建东 曹露茜 明春宏 吴婷 刘澎 陈勃 孙梅芳 易慧芳 杨丽萍等
景观面积：380 000 m²

## ▶ Overview 项目概况

The project is located next to Zhengshui River, Zhengxiang District, Hengyang, with superior natural conditions and environment. With suburb scenery to its north, Hengyang municipal government, Zhengxiang District government and newly built upscale communities in recent years to its south, bustling old town to the east, it will be the most beautiful riverside landscape avenue after completed.

项目位于湖南省衡阳市蒸湘区蒸水河边，自然条件优越，环境优美，北看城郊美景，南临衡阳市政府和蒸湘区政府以及近几年新建成的高端社区，东接繁华老城区。该项目建成后将成为最美丽的临江景观大道。

Site Plan 总平面图

### Design Concept 设计理念

With the rapid development of urbanization, traditional civilization being replaced by grass root and fast food culture, the legend of Zhengshui Thousand Year has faded, and the cultural character of an ancient city for thousands of years are covered by enforced concrete. The designers' mission is not only to build a park, but also pass on the fading memory of the land. Thus they propose a theme "Yancheng Memory Corridor", interpreted through three aspects: historic tales, humane custom and native landscape. It represents the historic humane landscape and native landscape in unique techniques.With the rapid development of urbanization, traditional civilization being replaced by grass root and fast food culture, the legend of Zhengshui Thousand Year has faded, and the cultural character of an ancient city for thousands of years are covered by enforced concrete. The designers' mission is not only to build a park, but also pass on the fading memory of the land. Thus they propose a theme "Yancheng Memory Corridor", interpreted through three aspects: historic tales, humane custom and native landscape. It represents the historic humane landscape and native landscape in unique techniques.

当下都市化的快速发展，传统文化被草根文化、快餐文化取代，千年蒸水的传说逐渐被淡忘，百世古城本应有的文化气息和底蕴也慢慢被钢筋水泥掩盖。设计师的使命不仅仅只是建设一个美丽的公园，还要传承这片土地正在流逝的记忆。设计师于是提出"雁城记忆长廊"的主题，从"历史故事、人文风俗和乡土景观"三个层面去诠释"记忆"，用独特的手法演绎、重现当地的历史人文景观和乡土景观。

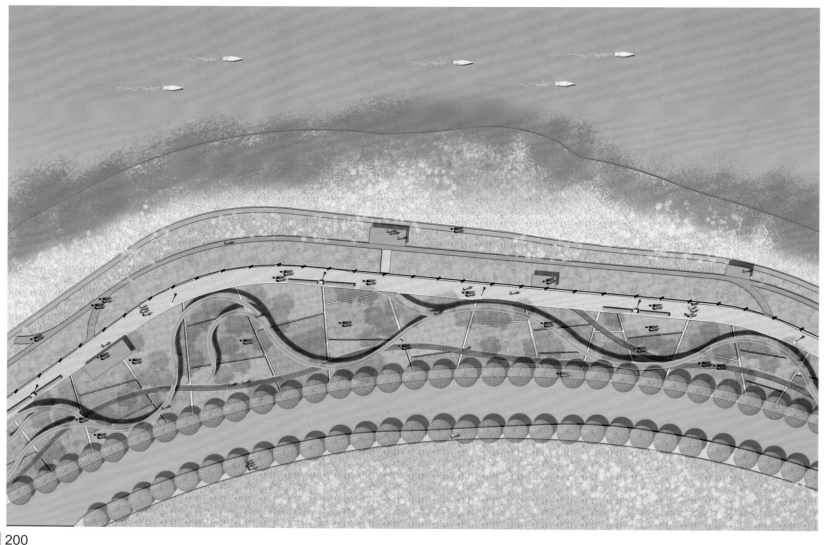

## Landscape Node 景观节点

The landscape design is expanded gradually with "Memory Corridor" as the clue, and five local characteristics as the breakthrough points: stone, paper, human, opera and field. The design adopts different landscape forms to present its five subjects and nodes: Stone Antiquities, Ancient Culture, Customs, Opera and Agriculture.

Stone Antiquities — Main Entrance Plaza and plaza for large events in the middle of riverside zone. It is inspired by the historic custom of carved stone record, i.e. Yuwang Stele, Stone Drum Academy, Huiyan Rock, manifesting people's favor of stone in Hengyang. As an imitation, designers set a large-scale landscape wall, with carved paintings to record the legend of Zhengshui River and folk paintings, showing the local unique customs and passing on Zhengxiang Culture.

Ancient Culture — Cai Lun is renowned in thousands of years for his invention of paper, which is the most proud item of Hengyang. Papermaking technology advances the development of ancient civilization, notes the history of Zhengxiang dated for thousands of years. The design with paper culture as subject reflects the development course of papermaking technology in a way of paper-cut silhouette.

Customs — the node with human as the subject, reflecting the local customs and lifestyle of Hengyang in dynamic and static ways, through theme sculptures "A Day in Hengyang" etc.

Opera — the regional Xiang Opera used to be familiar and popular while it is fading away. In order to present the character f Hengyang Xiang Opera, the designers set a node of opera culture. An open-air stage in the middle is used for daily practice by citizens or opera fans, or for performance in festivals. Some opera sculptures are positioned in the stage edge.

Agriculture — a landscape node presenting agriculture. Farming, rustic charm and landscape are the subject and characteristic of the node. Garden, soil, scarecrow, and bullfrog calls are not only memory of childhood but a peaceful land out of urban life.

景观设计以"记忆长廊"为线索，逐步展开，分别选取最具当地特色的五个点：石、纸、人、戏、田为切入点，概括为"衡韵"、"古韵"、"乐韵"、"曲韵"和"野韵"五个主题，分设五个节点，用不同的景观形式表现主题。

衡韵——滨江带中间的主入口广场、大型活动广场。主题来源于过往石刻记录的习俗，从禹王碑、石鼓书院到回雁石，无一不流露出衡阳人对石的钟爱。设计师效仿先人石刻记录的形式，设计大型石壁景墙，以雕刻画的形式记录蒸水河的传说及民间画，表现当地特有的民俗民风，传承蒸湘文化。

古韵——纸文化一直是衡阳人最引以为豪的，蔡伦是后世流传千古的英雄，造纸术的出现推动了古文明的进程，记录了蒸湘的百世繁荣。以纸文化为主题，用剪影的形式反映造纸技术的发展历程。

乐韵——以人为主题的节点，设置主题雕塑如"衡阳人的一天"等，反映当地民俗民生。并设置儿童活动和极限运动场地，给周边居民和游客提供游乐活动。从静态和动态两方面来反映衡阳人的生活。

曲韵——昔日曾经耳熟能详的地方湘剧，逐渐淡出茶楼戏台。于是设计师设计这个表现戏曲文化的节点，中间为露天大戏台，为市民或戏曲爱好者日常练习所用，也可于节庆日表演，边缘设置一些戏曲的雕塑小品，反映衡阳湘剧的特点。

野韵——代表农耕文化的景观节点，农耕、野趣、乡土景观是整个节点的主题和特色，田园、泥土、稻草人，还有牛蛙的叫声不再只存活在儿时的记忆里，而是成为存在于这片城市生活中的乡土新景观、繁华闹市里的一方净土。

# PUBLIC LANDSCAPE 公共景观

## MODERN STYLE 现代风格

### KEY WORDS 关键词

LANDSCAPE AXIS 景观轴线

LOCAL FEATURE 地域特色

LANDSCAPE NODE 景观节点

Location: Qian'an, Hebei
Client: Qian'an Construction Bureau
Landscape Design: ABJ Landscape Architecture & Urban Design Pty., Co., Ltd.

项目地点：河北省迁安市
客　　户：河北省迁安市建设局
景观设计：澳斯派克（北京）景观规划设计有限公司

# Hebei Qian'an City Huangtaishan Park
## 河北省迁安市黄台山公园

### FEATURES 项目亮点

The aim is to restore the hill, scarred by excavations and former habitation to its pristine value, and in this way to both complete the park and to remake its identity.

项目以恢复因挖掘和人类居住而被破坏的山体为目标，同时给新建成的公园创造特色，成为迁安市新形象的一个重要组成部分。

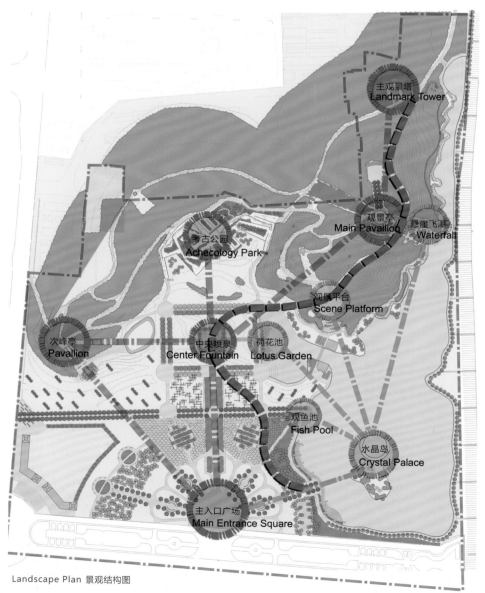

Landscape Plan 景观结构图

### Overview 项目概况

Huangtai shan Park is located in Qian'an of Hebei Province. It has been partially completed (Phase I), based on a master plan that has guided the initial direction. The ambition of this design exercise, is to retain the existing built landscape and to seek a natural balance between the man made and heavily occupied northern section of the park, and the remnant hilly, southern section of the park.

项目位于河北省迁安市，黄台山公园需要保留已经建成的（一期）景观工程，并寻求这些聚集于公园北部的人造物和位于南部的山体之间的平衡。

### Design Goal 设计目标

The aim therefore is to restore the hill, scarred by excavations and former habitation to its pristine value, and in this way to both complete the park and to remake its identity. In achieving this a complete experience in provided to the people of the City.

设计目的就是重塑被破坏的山体，还其一个质朴的本来面目，同时给新建成的公园创造特色，成为迁安市的新形象的一个重要组成部分。

## Design Ideas 设计思路

First, it emphasizes the advantaged location and chooses the site according to traditional philosophy of "Feng Shui" and "Long Mai". At the same time, the landscape design is ecology-oriented. As an addition, it has used geometry and axes in the design, which well shows the respect to the local history, culture and industry.

黄台山公园的设计思想由以下四个部分有机组成：首先是强调项目的区位优势，借鉴传统规划中的"风水"、"龙脉"的基地选择；同时将生态学很好地融入到景观的设计当中；此外还将几何学与轴线广泛地运用到设计之中去；整个设计很好地尊重了当地的历史、文化和工业。

# PUBLIC LANDSCAPE 公共景观

## MODERN STYLE 现代风格

# Linyi International Sculpture Park
## 临沂国际雕塑公园

**KEY WORDS** 关键词

ECOLOGICAL CHARACTER 生态特质

THEMED SPACE 主题空间

LANDSCAPE NODE 景观节点

## FEATURES 项目亮点

The design enhances the functions of the points, to enrich the fixed site spaces and bring abundant experience.

设计充分发挥点的作用，点即场所，丰富既定场所的空间，使其带给人们丰富的体验。

Location: Linyi, Shangdong
Landscape Design: Beijing Guanzhu Institute of Landscape Planning & Design
Chief Designer: Kong Xiangwei
Design Team: Li Guodong, Wang Lingling, Cui Jixiao, Han Shuo, Zhang Shuang
Landscape Area: 400,000 m²

项目地点：山东省临沂市
景观设计：北京观筑景观规划设计院
主创设计师：孔祥伟
设计团队：李国栋 王玲玲 崔继晓 韩朔 张爽
景观面积：400 000 m²

Site Plan 总平面图

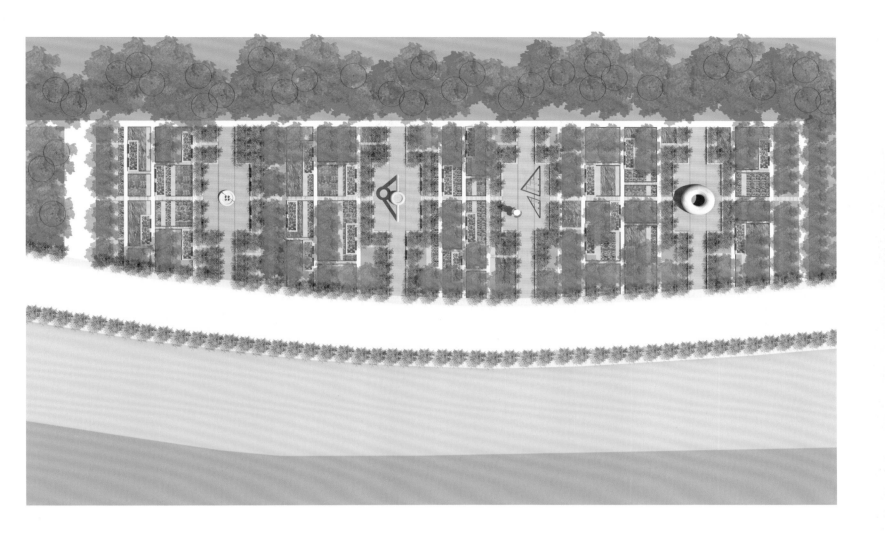

## Overview 项目概况

The project is located in the central area of urban new district, surrounded by river in three sides, a bathing beach to its west side while a bar street and TV Tower to its south. Its master plane covers the central zone and the belt zone in L shape. The design of central zone with nature and ecology as design core is composed of slopes, wetlands, water system, series theme spaces in point mode and a corridor throughout the zone. L-shape belt zone is mainly a understory sculpture zone, a public space with high anticipation composed of a series of theme parks and point buildings, understory green land.

公园所处位置为山东临沂城市新区中心区，三面环河，其西侧为沙滩浴场，南侧为商业酒吧街及电视塔。项目的总体布局包括中心园区和L型带状园区。中心园区以自然和生态为设计核心，由坡地、湿地、水系、系列点式主题空间和一条贯穿园区的廊桥构成。L型带状园区主要为林下雕塑参与区，表现为高度参与性的活跃公共空间，由系列主题园和点式建筑及林下绿地构成。

## Design Concept 设计理念

A sculpture park has a balance problem among art, ecology and humane which are also three basic questions Linyi Sculpture Park has to solve. The design use rhyme control to solve the problem among three items. It is a composition of key experience points connected with each other through lines to realize continuity and integrity. It enhances the functions of the points, to enrich the fixed site spaces and bring abundant experience.

雕塑公园需要解决艺术、生态和人性的三者平衡问题，这也是临沂国际雕塑公园需要解决的三个基本命题。设计利用控制节奏的方式，通过构图，布局关键的体验点，再运用线连接不同的景点，实现连续性和完整性。同时，充分发挥点的作用，点即场所，丰富既定场所的空间，使其带给人们丰富的体验。

### ▶ Landscape Node  景观节点

A curved gallery bridge links up six Lantern-theme sites in the central zone of the park, which is the solution to the anticipation problem and art express problem. The curved bridge with minimalism is built on most simple and basic structure, satisfying the security demand. The Lantern points are apt to be rich and charming for experience whose shapes and lighting are the medium of art expression. Central zone is the major sculpture exhibition zone, and the environment is set as the background of sculptures. The traffic and vertical system are arranged centralized the sculptures.

在公园中心区，利用一条曲线的路径兼廊桥串起六个"灯笼"主题场地。一条曲线加六个点解决中心园区的参与问题，以及艺术表达问题。曲桥利用极简主义方式，从最基本的结构出发，符合安全的需求；而点状"灯笼"的设计则趋向丰富性，以增加体验的魅力，"灯笼"的造型以及光作为艺术表达的媒介。中心园区是雕塑的主要展区，环境亦作为雕塑的背景存在，交通及竖向系统也围绕着雕塑展开。

L-shape belt zone also adopts the design of point-line-plane. The stone arrays in the plaza forms the sculpture space, decorated with exuberant native plants, leading to a strong contrast with the modern elements in the site. Understory sculpture zone is located in the broad part of the belt zone, whose themed sculptures with great sense of anticipation are connected through a path. Plants in understory sculpture zone are basically herbaceous plants and weeds in the way of brocade. A series of site are developed around the sculptures, generating urban vitality.

L型带状园区也同样利用点线面的对话，台地广场中的石阵形成雕塑空间，石阵中种以生命力旺盛的乡土植物，与场地中的现代元素形成强烈的对比。林下雕塑区，位于带状公园宽阔地带，一条林下路径连接了系列参与性极强的主题雕塑。林下雕塑区的植物基底采用织锦的方式，运用多种乡土草本植物及野草，作为林下的基础植物。系列场所围绕参与性雕塑展开，备具城市活力。

# PUBLIC LANDSCAPE
## 公共景观

## MODERN STYLE
## 现代风格

### KEY WORDS 关键词

**LANDSCAPE NODE**
景观节点

**LANDSCAPE CHARACTERISTIC**
景观特色

**WATERFRONT LANDSCAPE**
滨水景观

Location: Huangshi, Hubei
Landscape Design: Shanghai WEME Landscape Engineering Co., Ltd.
Total Planning Area: 160,000 m²

项目地点：湖北省黄石市
景观设计：上海唯美景观工程设计有限公司
总规划面积：160 000 m²

# Landscape Project of Huangshi Magnetic Lake
## 黄石磁湖湖景工程

### FEATURES 项目亮点

The whole design scheme focuses on the coherent terrain context on whole, emphasizing on the functions of leisure and the viewing, and highlighting the water-by characteristic and landscape effect.

整个设计方案在总体构思上注重地形的脉络连贯，强调休闲和观景功能，突出亲水性和景观效果。

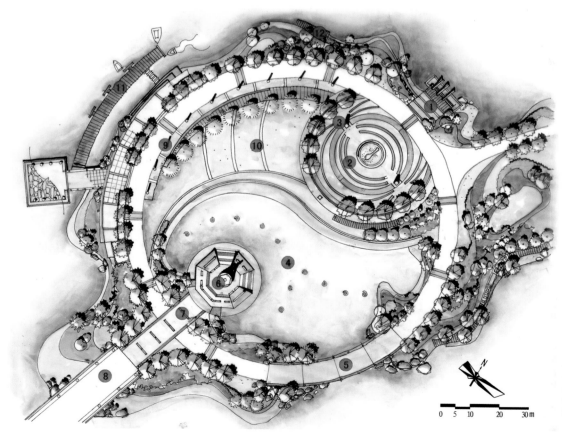

1. 磁湖司南
2. 磁湖之源——水罗盘
3. 景观磁柱
4. 磁湖喷泉
5. 鱼跃龙门
6. 秋月之塔
7. 特色磁极铺装
8. 磁堤
9. 特色景观灯柱
10. 阳光草坡
11. 溯源码头
12. 亲水栈道

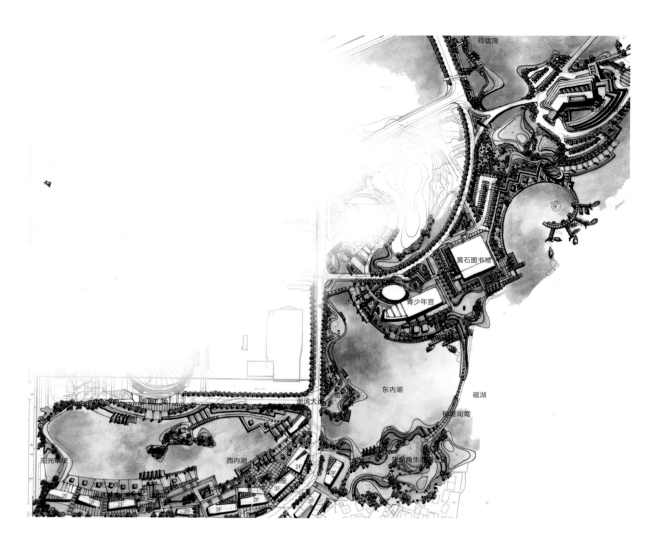

### Overview 项目概况

The project is located in Tuanchengshan Development Zone, and it is the key point of the project. Together with the People's Square, they form the core of the development zone landscape. The project base looks like a strip around the west of Magnetic Lake, and it faces Guilin North Road in the west, Xinqu First Road in the south, and is surrounded by Tuanchengshan of Magnetic Lake in the east and north. Water around the project base is part of Huangshi Magnetic Lake with an area of 178,168 m², and both its two sides can be used for green land construction with an area of 90,400 m².

黄石磁湖湖景工程位于湖北省黄石市团城山开发区，是湖景工程的重中之重，它与人民广场一起，组成开发区景观的核心。基地呈带状，环绕在磁湖西面湖畔，西临桂林北路，南临新区一路，东面及北面被磁湖中的团城山所包围。基地周边水域是黄石磁湖的一部分，面积为 178 168 m²，两边可用作绿地建设的绿地面积为 90 400 m²。

### Design Concept 设计理念

The whole design scheme focuses on the coherent terrain context on whole, emphasizing on the functions of leisure and the viewing, and highlighting the water-by characteristic and landscape effect. It also stresses to protect and regulate the natural environment on the basis of the combination of economics and aesthetics, and pays attention to the interaction between the landscape and people and the continuity of the activity project to reflect the characteristics of the city.

整个设计方案在总体构思上注重地形的脉络连贯，强调休闲和观景功能，突出亲水性和景观效果。方案还强调在充分结合经济学和美学的基础上保护和整治自然环境，并注重景观与人的互动性及活动项目的连续性，以体现城市特性。

### ▶ Landscape Features 景观特色

For the overall image and space processing, the project design is aimed to create affinity for the urban waterfront space in a pleasant way. Both on the aspects of the plants and space separation, the landscape project has beautiful scenery of the West Lake in different seasons, and the ornament style bears unusual exotic atmosphere as well.

The project design arranges different functional areas in linear according to the space sequence, thus creating a landscape area with different characteristics and easy to draw people's attention; the entire waterfront landscape seeks changes in the uniform and harmony in the comparison with distinct levels and rich rhythm and rhyme, forming a natural and interesting three-dimensional landscape and constituting a city life filled with natural humanity.

　　在整体形象和空间处理上，设计旨在通过宜人的尺度创造具有亲和力的城市滨水空间。景观在种植与空间分隔上具有西湖的四时美景，同时小品式样也融合了与众不同的异域情调。

　　设计把不同的功能场所，按序列沿线形空间展开，以创造具有不同特色却又富于参与性的景观区，使得整个滨水景观在统一中求变化、对比中求和谐，既层次分明，又富有节奏和韵律，形成一幅具有自然之趣和自然之理的立体山水园林画卷，构成一条弥漫着自然人文气息的城市生活走廊。

# PUBLIC LANDSCAPE 公共景观

## COMPREHENSIVE STYLE 综合风格

### KEY WORDS 关键词

PLANT LANDSCAPING
植物造景

SOFT LANDSCAPE
软景观

LANDSCAPE NODE
景观节点

Location: Shenzhen, Guangdong
Developer: OCT Group
Landscape Design: Shenzhen Joco Landscape

项目地点：广东省深圳市
开 发 商：华侨城集团
景观设计：深圳市筑奥景观建筑设计有限公司

# Shenzhen OCT East
深圳东部华侨城

## FEATURES 项目亮点

Tea Stream Resort Valley embodies entertainment and holiday tourism culture combining Chinese with western culture and integrates the customs of western mountainous towns, romance of wetland and flowers sea, and Zen & tea ceremonial culture and grace of Lingnan tea field.

茶溪谷体现了中西文化交融的休闲度假旅游文化，融合了西方山地小镇的风情、湿地花海的浪漫、茶禅文化的恬淡和岭南茶田的幽雅。

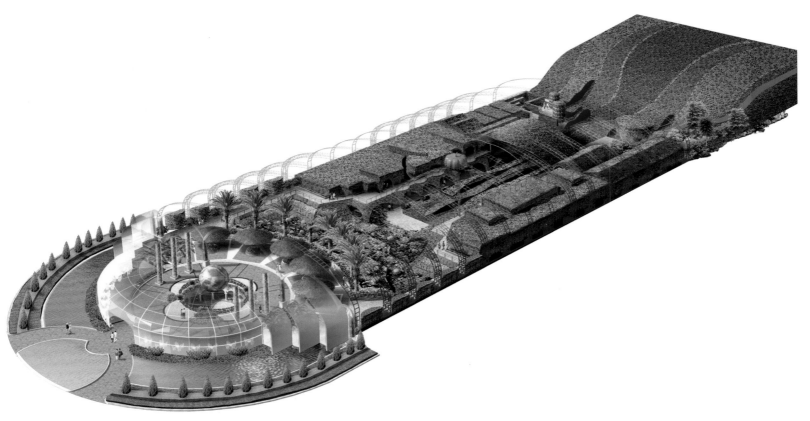

### Overview 项目概况

Located at Shenzhen OCT East, this project consists of Tea Stream Resort Valley, Four Seasons Plant Museum, Orchid Garden, Grand Canyon and Great Tsunami, etc. It is located in the tropical where there is ample sunlight, moist air by the sea, abundant wetland, canyon and lake resources and favorable natural environment. Nowadays, OCT East has become a gleaming city card of Guangdong-Hongkong-Macau, and is known as shining pearl in Shenzhen Bay.

项目位于深圳市东部华侨城,包括茶溪谷、四季植物馆、兰花园、大峡谷丛林穿梭、大海啸等。项目地处热带,阳光充足,依海而建,空气湿润,湿地、峡谷、湖泊资源丰富,有着得天独厚的自然环境。现在,东部华侨城业已成为粤澳港地区一张流光溢彩的城市名片,被誉为深圳湾畔的一颗闪亮明珠。

### Design Concept 设计理念

The design and planning depends on such environmental resources as the vegetation, brooks, lakes, canyons to reserve, improve and optimize the existing ecological environment, draws on development concept, operational modes and landscape planting methods of foreign ecological tourism areas and arranges ecological tourism scenic spots and projects of different types painstakingly in light of planning of tourism theme. Enhance ecological concept, stick to sustainable development, and pay attention to development potential in the industry to expand to the deep levels.

设计理念上,设计师一方面强调生态理念,坚持可持续发展的理念,借助滨海山地自然空间,依托植被、溪流、湖泊、峡谷等环境资源,对现有的生态环境进行保留、提升、优化;另一方面,借鉴国外生态旅游区的开发理念、经营模式和景观造林手法,结合旅游主题策划,精心布置多种类型的生态旅游景点和项目。

### Plant Landscape 植物造景

Tea Stream Resort Valley embodies entertainment and holiday tourism culture combining Chinese and western culture with the main elements of "tea, meditation, flowers and bamboos" and integrates the customs of western mountainous towns, romance of wetland and flowers seas, and combination of tea and mediation culture and grace of tea field of Lingnan.

Characterized by the sea of flowers and earth art, Orchid Garden is a wetland experience area that integrates sightseeing, science popularization education outdoor amusement with unusual sports. Visitors may meander through the labyrinth and gawk at the hundreds of bursting blooms in the orchid garden.

Concise and orderly wooden gallery frame are used to divide the space in Four Seasons Plant Museum, expressing a great charm of Chinese classical gardens. In addition, there are many landscape features, such as cane chair at the end of the museum, fountain near the footpath, large simulation birdcage at the road corner. What's more, copper-colored water wheel, juxtaposed sharp bamboo tube, native wood cross-sectional stitching scene wall and orchid in palm shape an artistic place that makes you contemplative.

In terms of verdurization, graceful trees, shrubs, flower and grasses distributed to create a dynamic and fashionable leisure space. In terms of plant arrangement, a variety of plants are planted, creating a natural, harmonious and well-arranged soft landscape.

茶溪谷体现了中西文化交融的休闲度假旅游文化，并兼有"茶、禅、花、竹"等主要元素，融合了西方山地小镇的风情、湿地花海的浪漫、茶禅文化的恬淡和岭南茶田的幽雅。

兰花园是以花海景观为载体，以大地艺术展示为特色，融合了观光、科普教育、户外游乐和特色运动于一体的湿地山野体验区。铺天盖地的鲜花、蜿蜒曲折的小径、美丽繁茂的观赏草，吸引人们进入兰花园雕塑区。整个兰花园雕塑以绿色植物组团为背景、亮丽的鲜花为点缀。

四季植物馆，设计师们利用简洁有序的木质廊架将空间分隔，极具中国古典园林的神韵。运用障景法打造出"庭院深深深几许"的赏游意境，框景法的运用又使游客仿佛置身于画中。馆内还设有众多特色小品，如在尽头设几把藤椅；在游步道上设一处喷泉；在园路拐角点缀几个大型仿真鸟笼等，让整体空间的形式里蕴含细节的连用感和序列感。景观细节上更是别出心裁，古铜色的转水轮、并置的尖竹筒、原生木横截面拼接景墙，以及附着的兰花，处处显露着令人沉思的景观意境。

项目绿化以乡土树种为主，选用树形优美的乔木，配以观花、观叶、灌木及草花，打造一种精致且充满活力、时尚的休闲景观。在植物配植上，尤其注重层次感，利用各式各样的植物打造出和谐自然、错落有致的软景景致。

# PUBLIC LANDSCAPE
## 公共景观

**MODERN STYLE**
现代风格

### KEY WORDS 关键词

THEMATIC CONCEPT
主题概念

CULTURAL CONNOTATION
文化内涵

LANDSCAPE NODE
景观节点

Location: Shenzhen, Guangdong
Client: Shenzhen China Merchants Group Shekou Industrial Park
Landscape Design: Shenzhen DongDa Landscape Design Co.,Ltd.
Total Planning Area: 48,000m²

项目地点：广东省深圳市
委托单位：深圳市招商局蛇口工业区
景观设计：深圳市东大景观设计有限公司
总规划面积：48 000 m²

# Shenzhen Shekou Taizi Road
## 深圳蛇口太子路

### FEATURES 项目亮点

In accordance with congenital condition of the site, building function and building requirements, this project creates a road environment of prominent thematic concept, high functionality and rich delight of life.

设计根据基地先天条件、建筑功能和建设要求等因素综合考虑，创造一个主题概念突出、功能性强和富有生活情趣的道路环境。

## Overview 项目概况

Located in the central area of Shekou, this project faces to the mountain and sea, boasting a unique geographical advantage. And the design of this project starts from Shuiwan Road to Shekou Ferry Terminal, which is a main road in Shekou that is adjacent to Taizi Plaza, Lizhi Park, Sea World and Shekou Ferry Terminal.

　　项目位于深圳市蛇口中心城区，坐山拥海，区位优势得天独厚。太子路本次设计范围为水湾路至蛇口码头段，与太子广场、荔枝公园、海上世界、蛇口码头毗邻，是蛇口区域重要的主干道。

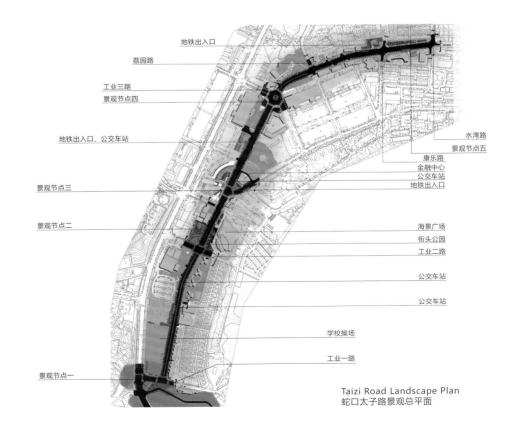

Taizi Road Landscape Plan
蛇口太子路景观总平面

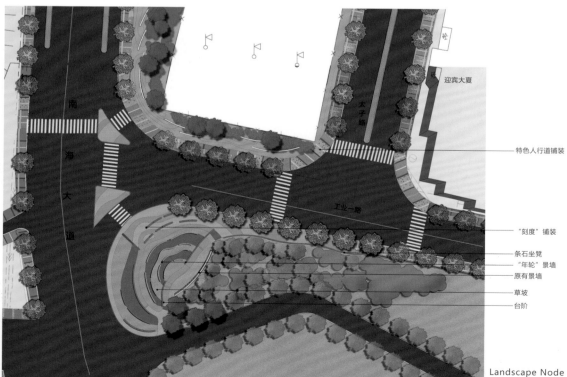

### Design Concept 设计理念

In accordance with congenital condition of the site, surrounding architectural style as well as local cultural connotation, this project creates a road environment of prominent thematic concept, high functionality and rich delight of life.

　　景观设计理念是将基地的先天条件、周遭建筑风格以及当地的文化内涵通盘考虑，创造一个主题概念突出、功能性强和富有生活情趣的道路环境。

Landscape Node Design - Site Plan
节点设计 - 平面布局图

## Landscape Node 景观节点

Designed in accordance with natural condition of the site, architectural features and construction requirements, it coordinates the relationship between the building, site and landscape, and is committed to transforming private space on both sides of the municipal road into public space, by dismantling the existing enclosed wall, it provides more public interacting and leisure space for the whole neighborhood. In addition, node space such as the five crossroads and roadside green space play the role of story-tellers, telling the development history of Shekou, which reflects the cultural connotation in a deeper sense.

设计根据基地先天条件、建筑功能和建设要求等因素综合考虑，全面协调建筑、场地及景观之间的关系，旨在将市政道路两侧的私有权属场地延伸纳入公共空间，拆除现状围合墙体保证整个街区拥有更多的公共场地空间，在景观设置上要求更具生活情趣，以促进公众参与成为可能。同时以蛇口发展历程为线索，以五个交叉路口或街头绿地形成的场地节点空间为载体，故事化地将蛇口的重点纪事融入整个街区，突出各节点肩负的不同主题概念，深层次展现蛇口的文化内涵。

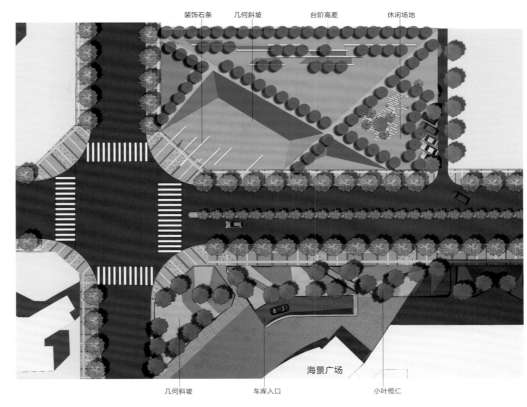

Landscape Node Design - Relaxing Space Plan 2
节点设计 - 休闲场地方案二

## PUBLIC LANDSCAPE 公共景观

## MODERN STYLE 现代风格

### KEY WORDS 关键词

LAKE
城市湖体

LANDSCAPE NODE
景观节点

PEDESTRIAN ROUTES
步道空间

Location: Tianjin, China
Client: City of Tianjin
Landscape Architecture: Atelier Dreiseitl GmbH
Architects: WLA, GMP, KSP, Riken Yamamoto, HHD, Callison, ECADI, TVSDESIGN
Engineers: Polyplan
Size: 900,000 m²

项目地点：中国天津市
项目委托：天津市政府
景观设计：德国戴水道设计公司
建筑顾问：WLA、GMP、KSP、Riken Yamamoto、HHD、Callison、ECADI、TVSDESIGN
工程顾问：Polyplan
项目面积：900 000 m²

# Tianjin Cultural Park
天津文化中心

## FEATURES 项目亮点

Cultural architecture integrates into the cultural and recreational area. Boulevard, green plants and artificial lake are set to form a comfortable cultural center.

项目设计的特色在于将城市建筑与文化休闲区域进行整合，运用林荫道、种植带以及人工湖等打造了一个舒适感极强的城市文化与休闲中心。

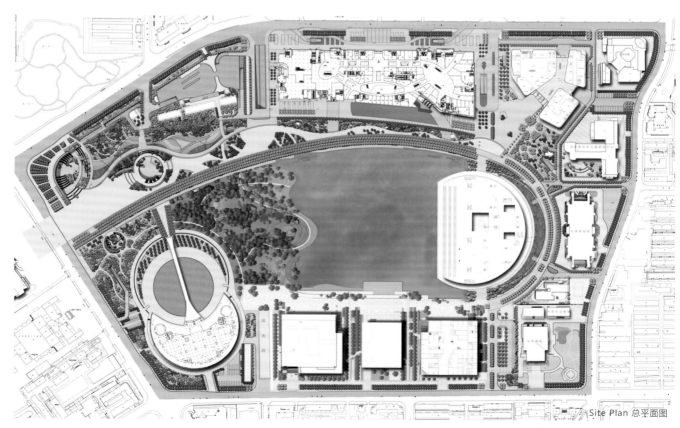
Site Plan 总平面图

### Overview 项目概况

Tianjin is one of China's top 5 cities, not just in size and population but also in terms of business investment. Located just half an hour south-east of Beijing by high-speed train, Tianjin is also close to the sea. The high-water table needs to be maintained to prevent seawater encroaching inland and the dry, harsh climate does not preclude flooding.

作为中国前五大城市之一，天津在城市面积、人口数量和商业投资方面都处于重要的位置。从地理区位看，天津位于首都北京的东南方向，乘坐半个小时的城际快轨即可到达。同时，天津也是一座港口城市，临近海洋，故需要保持地下高水位以防止海水渗透入内陆，使得土壤盐碱化，而干旱酷热的气候也难免洪水泛滥。

## Design Description  设计说明

In the design of the new cultural district between the new opera house and existing city hall, a main goal was to increase outdoor comfort and create dynamic, social pedestrian routes. The lake waterfront is aesthetic with dramatic views to the opera house and exciting Museum, gallery and library frontage. Avenues of trees and planting shield the waterfront from the cold Mongolian winds while at the same time storing water for irrigation.

The lake is a stormwater feature, a balancing pond which can effortlessly handle a 1 in 10 year storm event and buffer a 1 in 100 storm event. Generous tree plantings link subsurface, decentralised retention trenches which feed the lake via a cleansing biotope. The urban lake has its own natural biology and reduces temperature extremes in Summer. The scenic beauty sets Tianjin Cultural Park as most outstanding new cultural architecture in the city.

设计一个位于现有市政厅和新建大剧院之间的新文化区域的主要目标，是要增强户外空间的舒适感，并创建一个富有活力的可供社交活动的步道空间。湖滨的水岸则因望向大剧院和形态优美的博物馆、美术馆和图书馆建筑的动态视觉效果而充满美感。林荫道和种植带的设计使得休闲宜人的滨水步道区域免受冷风的侵袭，同时在功能上可储存雨水用于种植灌溉。

人工湖是一项雨洪设施，调蓄设计可以从容应对十年一遇的大雨侵袭，对百年一遇的暴雨也能起到缓冲作用。大量的树木种植与地下分散的滞留沟渠相连接，同时通过植物净化群落补给湖体用水。此城市湖拥有其自身自然式的生物特征，有助于缓解夏季的极端酷热。天津文化中心以其自身优美的风景成为了天津城市最亮丽的新文化建筑。